A WALK THROUGH THE VALLEY OF DEATH

A Soldier's Story

Dennis J. Fernandez, Sr.

with Sandra MG. Fernandez and Roberto Leal

Copyright © 2017 by Dennis J. Fernandez, Sr.

All rights reserved. No part of this book may be reproduced or used in any manner without written permission of the copyright owner except for the use of quotations in a book review.

DEDICATION

I would like to dedicate this book to the 58,320 Heroes and POW's/MIA's who gave their life for their Country, especially Jordan and Billy. I would also like to dedicate this book to the men/friends whom I had the honor to serve with in the rice paddies, jungles, tunnels, and combat. We ARE "Brothers". You will always be in my heart and in my thoughts.

The "War" took us in as boys and made us men!

82nd - Airborne "All the Way"!

INTRODUCTION

I had been meeting with a group of guys who had an idea about a monument for San Jose veterans of the Vietnam War. Mostly, they were guys just like me. We all came from working-class backgrounds. We all had come-of-age during the turbulent Sixties. We all had to deal with the draft at age 18. We all had served in "Nam" and experienced combat, survived and managed to come back home. We all had brought back memories of that war that still plagued us. We all were members of an exclusive club: Vietnam War combat vets. In many ways we were mirror images of each other.

When I came back home from Vietnam, I found that putting it all behind me was-- and still is-- a continuing struggle. It's hard to talk about it to anyone who hasn't lived through it. People are naturally curious about men in combat and always want to know, *"What was it like over there?"* It wasn't until several years ago when I was diagnosed that I was able to put a name to my ongoing struggle: PTSD (Post Traumatic Stress Disorder). I am often asked, "What does it feel like and what triggers it?"

Herodotus, a Greek Historian, 490 BC is credited in recording the first PTSD episode of a Greek soldier, who suffered from instant, temporary blindness, following a ferocious battle with the Persians.

The all too familiar sounds of rotating helicopter blades, oppressive sensation of intense heat, or exploding fireworks displace me temporarily and I'm back in Nam.

However, it's only a fleeting moment and I can bring myself back to the present. It's all about dark, shapeless emotions of anxiety, anticipation and the feeling that "something" is out there and I need to be alert. Unfortunately, sometimes I can be my own worst enemy. Drinking socially can lapse into overindulgence, taking me across the threshold into a full-blown PTSD episode.

Getting together with my mirror images and sharing combat stories helped some, but not enough. The PTSD definitely contributed to my two failed marriages. Fortunately, it didn't completely ruin or end my life like it has for so many other Vietnam Vets. Many Vets came back from Nam terribly and fatally damaged. Too many wound up homeless on the street, alcoholics, drug addicts, or dead from suicides. There's no war memorial for them. I was told that writing down my experiences in Vietnam might be a cathartic process for me. Frankly, writing about it is often just as hard and painful as talking about it. My attempts at writing about my recollections of Vietnam often turned into extended late night drinking bouts with Patron's tequila and Miller Hi-Life beer. I'm not a writer. Putting all this down on paper has not been easy for me.

I once heard that there is a timeless, classic way to begin any story. If you use these words, I was told, you open the door that takes you on a journey down the yellow brick road, on a magic carpet ride, or to the steaming hot, dangerous jungles of Vietnam. Just use these four magical words. *"Once upon a time…"*

Once upon a time, I was the All-American kid living a happy existence in San Jose, California.

Once upon a time, I was a boy who went into the Army and got sent to fight a war in a distant foreign land.

Once upon a time, I was a man walking point and leading a squad of soldiers on several dangerous combat missions.

Once upon a time, I was a tunnel rat and found myself crawling on my hands and knees with only a flashlight and a .45 caliber pistol searching for an invisible enemy hidden in a dark, claustrophobic maze.

Once upon a time, I served with some combat hardened warriors and earned their respect.

Once upon a time...they called me "Smooth."

CHAPTER ONE

Before San Jose and the Santa Clara Valley became the epicenter of Silicon Valley, it was the place Dionne Warwick sang about in "Do You Know the Way to San Jose". The year was 1947, and San Jose was more well-known for its fertile topsoil and abundant variety of fruits and vegetables. It was a valley dotted with orchards and canneries, not high tech start-ups and Starbucks.

I was the youngest of four children born to Lawrence and Gladys Fernandez. My Dad was a fifth generation Californian of Spanish, American-Indian descent. Mom came from a Scotch-Irish family that was originally from Missouri. It was ballroom dancing that brought these two very different people together.

The Coconut Grove is located on the world famous Santa Cruz Boardwalk right along the shores of the Pacific Ocean. You have to drive about 17 miles from San Jose over the mountains to get to Santa Cruz. But that didn't deter folks who loved to ballroom dance from gathering there and enjoying themselves. Mom was very impressed when she saw Dad out on the dance floor and sought him out as a dance partner. Soon, they were meeting regularly at the Coconut Grove and gliding across the floor together.

The Cocoanut Grove, Santa Cruz, California

A romance ensued and soon thereafter they were married. Neither one of my folks graduated from high school; Dad worked as a roofer and Mom was a homemaker, but also worked at the Dole Cannery during the summer months. Back in those days, if you were willing to roll up your sleeves, get a little dirty and put in long hours, you could make enough money to support yourself and even raise a family. My parents instilled in me, as well as my older siblings; Fred, Patrick and Nancy, the ethic of hard work and personal responsibility. My parents were part of a generation that was influenced by The Depression. They had to struggle and fight for everything they had. These were working-class people who labored hard, played hard, and yes, drank hard.

Picture of San Jose circa 1943

In 1943, my folks bought a brand new, 900 square foot home located on the other side of the tracks, literally. It was a modest two-bedroom house. I shared a bedroom with my two older brothers and sister. We were a traditional working-class family and never felt under-privileged or disadvantaged; we did the best with what we had. Although we didn't have a lot of "stuff", we were none the worse for wear. Like Bob Dylan said, "When you ain't got nothing, you got nothing to lose". I put those hard-working values my folks had instilled in me to work at the ripe old age of 12, when I started my own landscaping business. I, along with a neighborhood buddy, mowed lawns. I used my hard-earned money to buy the essentials of most 12 year old boys: sports equipment, marbles, Nehi grape soda and candy. I loved sports. During my school years, playing sports whenever I had the opportunity consumed my mind, body and spirit.

The Sixties was a time of great social change and cultural unrest in the United States. We lost our President, John F. Kennedy, in 1963. The Beatles landed on our shores and appeared on the Ed Sullivan Show the following year. In 1964 and 1965, President Lyndon Johnson signed into law sweeping legislation: Medicare, The Civil Rights Act, and The Voting Rights Act, as well as declaring a War on Poverty. Johnson also escalated our involvement in another little war in a part of the world initially known as Indochina.

However, this much is clear: 1965 was not the Sixties. That iconic period of time didn't really get started until the Summer of Love in 1967. In 1965, things were still relatively conservative and tame. We still wore our hair short and neat. And even though the British Invasion was gaining momentum by 1965, the number one selling record that year, according to Billboard, wasn't The Beatles' "Help," Bob Dylan's "Like a Rolling Stone," The Temptations' "My Girl," or even the Rolling Stones emblematic rock anthem of the ages: "I Can't Get No Satisfaction." No, in 1965 the number one selling record in the United States was "Wooly Bully" by Sam the Sham and the Pharaohs. When the Voyager spacecraft was launched in 1977, several phonograph records were sent along so that any other intelligent beings out there in the cosmos could hear the kind of music Earthlings were creating back in our neck of the universe.

During this time, my high school, San Jose High, was a fantastic blend of many divergent cultures: Hispanic, African-American, Asian, and Italian. We went to school together, played sports together, and partied together. The undeniable prevailing atmosphere of racial harmony was uniquely present at that particular time. This was the historical and cultural backdrop that I grew up with and shaped my outlook of life.

At San Jose High, I lettered in wrestling, track and field, and basketball. I definitely saw myself as an above average athlete, but far from the college level. However, basketball was, and still is my passion. Having fun and partying was another priority of mine in high school. Back in 1965, San Jose was an ideal place to party on the weekends. San Jose State College was ranked as the #1 party school in the nation! Go figure! Imagine little San Jose State upstaging giants like UCLA, USC and UC Berkeley. There was always some kind of party going on at a dorm or apartment complex where college students lived. Although I was only a high school senior, I never had any problem getting in and mixing with the college kids. The music at these get-togethers tended to lean towards the more typical party music; "Louie, Louie" by The Kingsmen, Beach Boys music, The Supremes, and The Beatles. On the other hand, the parties that sprung up around 13th and Julian streets tended to have more of a Latino and African-American ethnic mix. You could drink beer, dance, and visit to the sounds of Jimmy Reed, James Brown, The Temptations, Aretha Franklin and Marvin Gaye.

One of the primary weekend rituals of high school life in the early '60s was "dragging the main." Before texting and social media, you connected with your teen peers by dragging the main. This entailed driving up and down a popular street to show off your car, and who may be riding in it. So naturally, it required having a set of wheels…a car. I had been driving an old '53 Chevy until one fateful Halloween night in my junior year. My friend Bobby Leal and I were driving around trying to track down two particularly attractive young ladies from our class. As I stopped at a corner, out of nowhere this idiot in a big Caddy slammed into us head-on and smashed up the front end of my car. Luckily nobody was hurt. But not only was the other driver totally at fault, he was uninsured and drunk! I got a '57 Chevy out of the settlement, thanks to my Dad negotiating the deal. In 1965, the '57 Chevy hadn't quite reached classic auto icon status. Back then it was just another eight year old car. But, it was a great car for me.

Dragging the main was immortalized in the 1973 George Lucas film "American Graffiti." A great deal of the action in "American Graffiti" takes place inside of cars with teenagers driving around and occasionally racing on city streets. It's a seminal coming-of-age film that neatly captures the California car culture of that era, as well as exploring the tough decisions many draft-age boys had to make as the storm clouds of the Vietnam War formed.

The route that encompassed "the main" was basically driving up First Street past St. James Park and the movie theaters, and then over and down Second Street. There were other tributaries that crisscrossed, but that was "the main". This was how you met girls, confronted rivals from other high schools, and found out where the best parties were happening that night. If we got hungry, there was cheap food at Tico's Tacos and the Burger Bar. If you needed beer, there was always a wino that was willing to buy you a couple of quarts of beer at Kelly's Liquors on Fourth Street.

Going out on dates was so inexpensive back then. The drive-in movies were only $1.50. After the movie, you could treat your date to burgers, fries and strawberry malt for a couple of dollars and change. Gas prices averaged around 31 cents a gallon. So the old '57 Chevy never went thirsty and was always ready for a drive. I earned money by working at a snack bar at my high school; I was happy to provide snacks for adults attending night school. I would go to practice, afterwards do my homework, and then head to work. I made enough money to finance my weekend escapades. Yeah, 1965 may not have been the *real* Sixties of legend and lore, but it was a great year for me. I was on top of the world and ready to conquer it. I was a decent looking kid, a good athlete with a set of wheels, and had a steady girlfriend. These were qualities and attributes so many of the young men in San Jose shared in common.

We are not about to send American boys 9 or 10 thousand miles away from home to do what Asian boys ought to be doing for themselves.
---President Lyndon B. Johnson

In April of 1965, I turned 18 and went down to the U.S. Post Office to register for the draft as required by law. The world I was sitting on top of was starting to spin in different grooves that would take my life in a very different direction.

Graduation Portrait, 1965

Prom photo of me, 1965

Graduation day photo of me and my cool '57 Chevy

CHAPTER TWO

We must walk consciously only part way toward our goal and then leap in the dark to our success ---Henry David Thoreau

It was around this time that I felt the urge and need to "leave the nest;" it was definitely time to spread my wings and find my own place to live. Shortly after my high school graduation, I packed my stuff and moved on with my life. I shared an apartment with a high school buddy of mine named Oscar Reyes. Oscar was an exceptional athlete, particularly in basketball, and went on to compete at the community college level. I had a summer job at the Dole Pineapple Cannery which helped pay for my portion of the rent, tuition and books. In those days, you had to know someone working in the cannery in order to get a job there. Fortunately for me, my Uncle Mario was a big shot mechanic at Dole, so getting a job was a piece of cake. I was also working for the City of San Jose as a recreation leader. Sometimes, I helped my Dad do side jobs roofing houses. Roofing is hot, hard, and dirty work. In the summertime, when Dad did these side jobs, the temperature on those roofs could reach 112 degrees! I have to admit though; roofing was not my cup of tea. One hot, summer day on the roof, my normally quiet father tapped me on the shoulder and suggested I go to college and focus on another occupation. I believe my father knew I had potential to be successful in life if I just kept my dreams and goals alive. This short, but significant conversation validated my aspiration to stay the course. Don't get me wrong, if it weren't for journeyman roofers like my dad, there would be a lot of leaky roofs, especially if I did the work. So off to college I went!

Oscar and I were enrolled at San Jose City College. My career goals were to become a teacher, coach basketball and football, and eventually an athletic director. I had lofty ambitions, but unfortunately my enthusiasm for attending class to reach those ambitions didn't always match. I was a young, 19 year old carefree kid having too much fun running and sliding down the halls of the college. There were too many parties to go to and football games to attend to bother with going to class.

I remember one football game San Jose City College played against San Francisco City College. The team from San Francisco had a running back that scored several touchdowns against us. Subsequently, this running back got transferred to USC where he went on to the win the Heisman Trophy. After a great NFL career, this running back from San Francisco City College, O. J. Simpson, was inducted into the NFL Hall of Fame. After the aforementioned game at San Jose City College, there was a dance and O.J. walked in with a young lady on each arm. He boasted for all to hear, "I just scored four touchdowns in the mud!" It was obvious even then that O.J. was well on his way to becoming not only a great NFL running back, but a famous sports celebrity. Who could have predicted that it would all end in such infamy and personal tragedy?I felt I was making a good decision to stay in school. College life was growing on me, and I was settling into a comfortable school routine. However, I was also beginning to hear about the U.S. involvement in Indochina, though I never considered that I would get drafted. I was learning that our involvement in Vietnam had actually started right after World War II; our commitment to continue our involvement in South Vietnam passed into the Eisenhower administration during the '50s. Under President Kennedy, we began to send military "advisers" to help the South Vietnamese fight the communist Viet Cong and their North Vietnamese allies. There were reports on TV about American military personnel getting killed in Vietnam. In those early college days, it all seemed so distant and removed from me, but the number of U.S. troops in South Vietnam was increasing at an alarming rate.

In 1963, President Kennedy had sent 15,000 military advisers to aid President Diem in his fight against the Viet Cong communists. After the Gulf of Tonkin Resolution in 1964, President Johnson had raised the troop level to 200,000. By the end of 1965, that number increased to 400,000, 1967 saw that level increase to almost 500,000 U.S. military personnel in South Vietnam.

All these facts, figures, and news stories about a country I probably couldn't find on a map about a conflict I didn't totally understand seemed like a boring history lecture to me. The idea of walking around in some far off jungle in Asia wasn't even the faintest blip on my radar screen. Unfortunately, Uncle Sam had different plans for me. He was going to give me a chance to learn all about Vietnam…up close and personal.

My first draft notice arrived in January 1966 with specific instructions to report to the Oakland Veterans Administration Hospital for my physical. Receiving that draft notice actually set me back a little. I thought to myself, "Wow, I actually got drafted. It really happened to *me!*" The Registrar of Selective Service was my very first stop. I took my draft notice there and informed them I was enrolled in college and was entitled by law to a deferment. Obtaining a college deferment was a common method of delaying or postponing military service via the draft. I left the Registrar's office with my deferment and a smile on my face. What a relief. I wasn't going into the military, let alone Vietnam!

September 1967 rolled around and there was Uncle Sam knocking at my door again inviting me to the Big Dance in Vietnam. I had received my second draft notice.

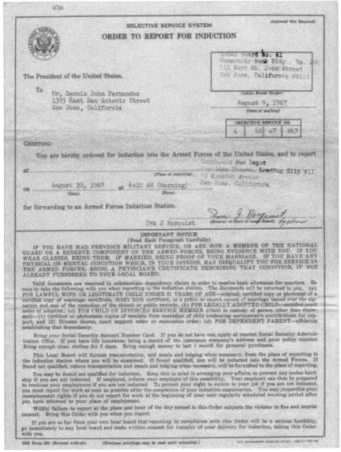

My draft notice

It's hard for me to explain how and why I chose to take the path that led me to serve my country. It was not a clear and distinct decision, but more like a feeling growing inside of me. I'm sure I was influenced by the long history of military service in my family. During World War II, my dad (he was too old to enlist and had 3 children) served his country by working in the Oakland Shipyards. He helped build Landing Craft similar to those used on the beaches of Normandy on D-Day. My uncle, Clarence Bond, was in the Navy stationed on one of the ships that was bombed by the Japanese in Pearl Harbor on December 7, 1941. He was a mechanical engineer and was caught below deck in the engine room when the attack hit his ship. He was up to his shoulders in water and oil for hours before being rescued. He is on a short list of Pearl Harbor Survivors. There was also Uncle Mack, who was in the Army during World War I; Uncle Mack later became a professional boxer and was inducted into the San Jose Boxing Hall of Fame. My two older brothers served in the military as well. The eldest, Fred Fernandez, was a sergeant in the

Fred Fernandez (brother), 1957

Marine Corps from 1954 to 1957. My other brother, Pat Fernandez, was in the Army from 1959 to 1961. Although the United States was involved in military operations at various hot spots around the globe at that time, it was fortunate neither one of my brothers saw any combat action.

It was time for me to follow in my proud family tradition of

Pat Fernandez (brother), 1959

military service. It was time for me to stand up and be counted. Instead of going back to the Registrar of Selective Service to ask for another deferment, I offered myself to the volunteer draft in October 1967. By volunteering to enter the military in October, I understood there would be a couple of months of boot camp followed by one last Christmas at home. I would then only have Advanced Infantry Training before I would have to go overseas to Vietnam, assuming those would be my orders. My mind was made up; I was going to Vietnam! I was 19. I was going to serve my country; I felt it was my duty. Despite the long tradition of military service in our family, all my siblings encouraged me to consider the option of going to Canada to avoid being sent to Vietnam. Did they know something I had not realized?

A Walk Through the Valley of Death

CHAPTER THREE

Nowadays, people tend to forget that during the Vietnam War era, draft-age young men had many military service options available to them. There were branches of the military that made it less likely you would land "in the bush" in Vietnam. Discussions at parties among other draft-age boys often revolved around this question, "What are you going to do if you get drafted?" There were several choices to choose from the military service menu. You could become a "Weekend Warrior" by joining the National Guard or Reserves. Although joining the National Guard or Reserves meant a six year commitment, active duty was only a couple of months out of the year and attending monthly meetings. If you were in the Guard or Reserves, the odds of being sent to Vietnam were slim to none. There were some of my high school buddies who had parents, family or friends who had contacts at the Selective Service Office who would put their names at the back of the file cabinet, thus "never" getting a draft notice. Many Air Force, Navy and Coast Guard personnel did tours of duty in Vietnam and other places in Southeast Asia during the Vietnam War. However, with very rare exceptions, the vast majority of enlisted airmen and sailors were never put directly in harm's way. They were "in the rear with the gear." But the principle drawback to these branches of the service was their four year term of enlistment. Four years? Four years to a 19-year old kid might as well have been 400 years. Four years was an eternity.

My sister Nancy, who is three years older than me, had always been my personal protector in grade school. Back then, I was a skinny little runt and sometimes got picked on by bigger kids. Nancy was always there to defend and stand up for me. Having grown up with two older brothers, she was a tough as nails tomboy. It would have been great to have her by my side in Vietnam.

My sister Nancy

I had no idea of what it meant to go off and fight in a war. Like all the young men of my generation, I bought into Hollywood's romanticized depiction of war. For me, John Wayne, in such classic war movies like "The Sands of Iwo Jima", "Back to Bataan", and "They Were Expendable" epitomized what being a soldier was all about. I would have been more impressed with John Wayne if he had served his country in WWII; many other superstar actors of that era served their time. I wasn't particularly interested or knowledgeable in the complicated history, politics and intrigue that got us involved in South Vietnam. My idea was simple: "I'll go over there, kick some ass, and come home." Oh, if only that were true.

October came too quickly. It was time to report to Oakland for my physical. By this time, I had moved back home with my folks until it was time to report for basic training. The physical essentially consisted of a team of doctors checking to see if I was breathing and had a pulse. After meeting those few requirements, I was declared "good to go." Shortly afterwards, I took the pledge to protect and defend the Constitution of the United States. I was now officially in the United States Army. One also needs to remember that most of us under 21 did not have the right to vote, but here we were off to fight in war for our country.

A Walk Through the Valley of Death

CHAPTER FOUR

My basic training began in October 1967 at Fort Lewis, Washington. Basic training lasted six weeks. During that six week period, one thing kept puzzling me, "What does the terrain and weather in the state of Washington have in common with Vietnam?" The answer was simple… nothing. The only thing Fort Lewis, Washington and Vietnam have in common is that they are both located on planet Earth.

Vietnam is in the hot, humid tropics where temperatures can reach 100 plus degrees. By comparison, during my training at Fort Lewis, we often experienced snow, sleet, freezing rain and temperatures that ranged between 25 and 40 degrees. Thankfully, after my six weeks of basic training, I went home for thirty days. I was able to celebrate one last Christmas with my family just as I had planned when I signed up for the volunteer draft.

On November 20, 1967, San Jose State College students demonstrated against the Dow Chemical Company, the maker of napalm. Police were sent in, but the students refused to disperse and several

My Military Photo, October 1967

protest leaders were arrested. The next day, students defied Governor Ronald Reagan's warning against further demonstrations and again staged an anti-Dow demonstration. Napalm was an acronym derived from naphthenic and palmist acids, whose salts were used to manufacture the jellied gasoline, napalm, which was used in flamethrowers and bombs. Napalm first came into widespread use during World War II, especially in flame throwers used to destroy entrenched Japanese positions in the Pacific War. It was also used extensively in aerial bombs during the Korean War against Chinese and North Korean entrenchments. The use of it in the Vietnam War was to kill vegetation in order to readily expose the enemy *Because of its horrible burning effects on the human body, many Americans considered it an especially cruel and barbaric weapon.* ---Archives and Oral History, University of Wisconsin.

Spartan Daily article at San Jose State College

In January 1968, I was back in Fort Lewis for Advanced Infantry Training (AIT). The 4th Infantry Division had been assigned to get us ready for

Me, April 1968

combat in Vietnam. The 4th had the dubious distinction of being the Army unit that had suffered the most casualties in the Vietnam War! I guess the Army, in their infinite wisdom, had figured we would learn from their "mistakes." Factoring in that our AIT took place in typically Northwest winter conditions of snow and rain, I was getting the queasy feeling in my stomach that I was not going to be prepared for combat in the jungles of Vietnam. The 4th Infantry didn't show us how to detect mines or booby traps. There was no instruction on how to go down into a tunnel armed only with a .45 caliber pistol and a flashlight, much less what to actually do when you entered one of these frightening black holes. We were not taught such important combat tactics like walking point or setting up an ambush. Hell, they didn't even show us how to handle the flesh eating ants, mosquitoes and leeches that were the bane of a grunt's existence in the "bush." Aside from being plagued by a sense of not being prepared for combat, I was also going

Me and mom, Gladys April 1968

through a sense of denial. Somewhere in the back of my mind-- and I believe in the minds of most of the guys in my unit at that time-- I really didn't believe we were going to be sent to Vietnam; maybe that's what I wanted to believe.

It wasn't exactly a slam dunk that if you got drafted into the Army or Marines that you were guaranteed a tour of duty in Vietnam. There were U.S. troops stationed all over the world at that time. Possible assignments included duty in South Korea, Okinawa, or maybe even Germany. Stateside duty was also a possibility. I'd even heard about soldiers during World War II who spent the entire conflict stationed at Fort Dix, New Jersey. But the dice didn't exactly roll in my favor. While 5% of the guys I was with in AIT went to parts unknown, I was in the "lucky" 95% that got orders for Vietnam. With an anxious sense of unknown expectations and a gnawing feeling of unpreparedness, we were sent home on a 30 day leave to await our departure overseas. During my leave at home, the anti-Vietnam War protests were beginning to heat up with red-hot intensity. On March 6, 1968, students staged another protest against Dow Chemical. Dow was the principal producer and supplier of napalm to U.S. combat forces in Vietnam. I would later come to appreciate the effects of dropping napalm on enemy positions. A few well-placed canisters of napalm would burn, disperse or kill a great number of the enemy. It also destroyed all the foliage thereby depriving them of effective cover. But aside from these distractions on TV news, my leave was fairly uneventful.

It was hard to tell what my parents were thinking at this time. Not much was said. My parents were quiet people. However, I'm sure they were very concerned that their youngest child was going off to fight in a war. Now that going to Vietnam was a certainty, my siblings urged me to reconsider the Canada option. Family conversations centered on me

Me, December 1967

Fred, Me, my dad Lawrence and Pat, December 1967

completely opting out and going AWOL (Absent With Out Leave). Truthfully, the thought of getting killed in a far off country, which I knew virtually nothing about, did not sit well with me. But the die had been cast. The decision had been made by me and Uncle Sam. I was going to Vietnam.

Me and my sister, Nancy, dancing at my farewell party.

Prior to leaving, my sister, Nancy, threw me and her brother in law a party. It was a nice gesture and very much appreciated. It was a great way to ease the tension prior to going to war. She has always been thoughtful and caring.

On April 4, 1968 my then fiancée drove me to Travis Air Force Base in my '57 Bel Air Chevy. It was a quiet ride to Travis. We didn't say much. Occasionally, the tears would well up in her eyes as she contemplated the certain danger that lay ahead for me. But to her credit, she handled it well. She exhibited a stoic strength that surely resulted from her strong,

Spanish and Italian heritage and conservative Catholic upbringing.

As I boarded the plane with 150 other naive soldiers bound for Vietnam... or as we ironically referred to it: Flower Land, there was a nagging question buzzing round in my head, "What the Hell am I doing?"

CHAPTER FIVE

When the plane taking us to Vietnam made a stop in Hawaii, I had a soul-searching moment. I asked myself, "Why am I going to fight in this war while people back home were protesting our involvement?" It was a faraway place that many people couldn't even find on a map. I had no interest in politics during my high school and city college days. Sports, partying and having fun were my primary interests in those days. However, I didn't exactly have my head in the sand regarding current events and the growing controversy over our involvement in Vietnam.

By 1968, the prevailing attitude in the nation and establishment media was supportive of our efforts in Vietnam. But, there was a growing vocal and often violent anti-war sentiment, particularly on college campuses. The dominant narrative I learned in my political science class was that we were assisting the democratic government of Vietnam; this narrative was echoed by the establishment media at that time. We were in the middle of the Cold War fighting the communist Red Menace of Russia and China who were vying for world dominance.

The Domino Theory, as developed during the Eisenhower Administration, asserted the belief that if Indochina fell to the communists that other countries in Southeast Asia, Malaysia, Indonesia and the Philippines would in turn-- like dominoes-- come under communist control. According to this theory, if communist aggression was not checked in Indochina, Hawaii would eventually be overrun, meaning that the communist threat could potentially threaten the U.S. Mainland. The Domino Theory dominated the foreign policy thinking in Southeast Asia through the Kennedy, Johnson and Nixon presidencies.

The principle of the Domino Theory was repeated over and over to us in basic and AIT; "If we lose Vietnam, the communists will eventually come ashore on the beaches of Hawaii." I realize now that it was all a form of brainwashing to get us mentally and physically prepared to face the enemy. I had bought into their hype!

The layover in Hawaii only lasted about 90 minutes. Nobody was allowed off the plane during the layover. Why not? We all looked at one another and laughed because we all knew why. The Army didn't want us to get off the plane in Hawaii because of their unspoken fear that many of the troops might go AWOL.

Finally on April 4, 1968 our troop plane was making its approach to land at Bien Hoa Air Base. But there was trouble brewing down below on the ground. Our captain informed us that the enemy was "waiting" for us near the landing strip. The plane lands abruptly. Our hearts are beating out of our chests! We are all scared and we are nervously waiting for the ramp to drop. What are we getting in to? We were told to "Get the hell out of the plane and haul ass to the cement bunkers!" It seemed like an eternity to reach those cement bunkers. We had been there only a few minutes and we were already taking mortar and rocket launcher fire! Inside the bunkers we were soaking wet from the scorching heat and oppressive humidity. We were all scared to death. There we were in the bunker, under heavy fire and we hadn't even been issued our weapons. We were totally defenseless. As we heard the mortars and rockets exploding around us, we all realized that if we were overrun, our only means of defending ourselves would be hand-to-hand combat. The temperature that spring day was an unbearable 100 plus degrees; my dirty uniform was thoroughly drenched with sweat. In my mind, I had arrived in the depths of Flower Land and the Devil himself, came to welcome me and drag me down to Hell.

Although not soon enough, we got the "all clear." We grabbed our gear and headed for our barracks. Luckily, there were only minor injuries. It is a well known fact that intense combat events can cause a soldier to crap his pants. After feeling defenseless under heavy rocket and mortar attack, unarmed and running for our lives, some of us needed a clean change of underwear.

Back in the USA, April 1968 ushered in what would prove to be a very turbulent and eventful summer. The very same day we arrived in Vietnam, we learned that James Earl Ray had shot and killed Martin Luther King, Jr. in Memphis, Tennessee. King's death set off a series of urban riots. The Black Panther Party emerged as a more aggressive and militant alternative to Martin Luther King's vision of a peaceful, non-violent Civil Rights struggle. Students all over the world were protesting oppression, racism and the Vietnam War. In places like Poland, Czechoslovakia, France, and Mexico, young people were going up against authority and often paying for it with their lives. Anti-war protests in the United States intensified to the point that following the Tet Offensive in January 1968, President Johnson announced to the nation on TV that he "will not seek and I will not accept the nomination of my party for another term as your President." From my vantage point in Vietnam in 1968, it seemed to me that President Johnson was jumping ship. Under his presidency, the conflict in Vietnam had become an American war, his war. And now that he was feeling the heat, he decided to get out of the kitchen.

With all the turmoil and discord at home and around the world, a sense of alienation came over me. Here I was, ready to lay down my life for my country, meanwhile back home the growing sentiment seemed to be "let's get out." My motto quickly became, "Cover your ass and the guy next to you and get back home." Eight days later on April 12, I turned 21 years old. Vietnam wasn't exactly where I had planned to celebrate, but I made the best of it. A good buddy brought me some welcome birthday cheer in the form of two special gifts: a six-pack of Pabst Blue Ribbon beer and a hand rolled marijuana joint the length and thickness of an expensive Havana cigar. Ah yes, it was a happy, high time in Flower Land that day. At dusk, while looking through the rifle outlets of our bunker, I observed a beautiful sunset. After that, I watched the bright red tracers that occur after 10 or 15 rounds of a machine gun burst, light up the evening sky.

Red Tracers in the Night Sky

The last of the "fireworks" came from the pink and white flares of the artillery guns. It was indeed a beautiful, chemically enhanced and scenic birthday. This was my first introduction to marijuana and the places it takes you.

In the following two weeks, we received intensive training from a group of battle-tested vets. These guys had all been in the thick of the action. After combat assignments in the field, they would re-emerge and train new troops, then, just as quickly, they would return to the action. Without a doubt, it was the best combat training I ever received. Far superior to anything I had received in basic training or AIT. These guys really prepared me for what lay ahead. They taught me invaluable skills vitally essential to succeed and survive in combat. Whether they were ordered to assist the newcomers or not, I was very appreciative of their training. They were my true mentors during my tour of duty in Vietnam. By the time they left, I had learned how to set up clay more mines, trip flares, locate and define different types of booby traps. I learned how to walk point, flank point, what to be aware of in the jungle and on the roads. They also taught me how to use a compass, read maps and call in artillery support. We even took time out to fire an M-79 (aka grenade launcher).

I finally felt prepared, as was humanly possible, for combat. Even down to minor details of using our water pills to prevent dysentery and taking anti-malaria pills on a weekly basis. They taught us very important survival skills.

Me with an M-79 (grenade launcher)

I was now a full-fledged grunt prepared to go into "the bush" (jungle) to hunt down "Charlie" (Viet Cong).

I was assigned to the 82nd Airborne Division, Company D, 1/505th Infantry. Our assignment was as a support unit for the 101st Airborne and the 3rd Marine Division. They were located in the northern part of South Vietnam just outside the ancient imperial capital city of Hué.

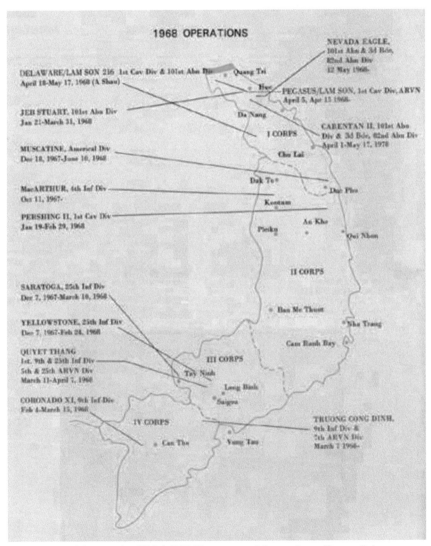

(Tour 365, USARV Returnee Magazine, Page 62)

CHAPTER SIX

In war, an infantryman goes through several phases of growth to mature into a combat ready soldier. First, you are a "cherry". In Army slang this referred to a soldier (officer or enlisted) who has just arrived at their first duty assignment after completion of training. But in the 82nd Airborne and other units in Nam, it specifically referred to a grunt that had not been in combat yet. The term "cherry" was an obvious wink to the "virgin" status of a G.I. not yet baptized in the bloody waters of combat. May 1, 1968, my May Day festivities were not the usual ribbons and dancing around the May poll, but instead it consisted of a time-honored combat rite of passage, I "popped my cherry."

It was time to put all of my training into action. Our company went out on patrol. On these missions someone is assigned to walk point. The grunt walking out in front of his unit is "the point man." Normally, he's several meters in front of the group. There are also guys "walking flank point." They are usually about 50 meters on either side of the company or platoon. Walking point is dangerous business. You need to be hyper-vigilant for the presence of snipers, unusual movements, trip wires, and anti-personnel mines, known as "bouncing Betty's." These devices are designed to inflict serious harm on the point man and to slow down troop movement. I've talked with some of my fellow Vietnam-era buddies who also walked point and asked which position they considered more "hazardous to your health:" front point or flank point. Traditionally, the guy taking point out in front of his unit was usually the first to encounter hostile fire. However, others who had walked flank point said that being so far off to either side of the unit made them feel more isolated and especially susceptible to ambush or snipers. Either way, we unanimously agreed that walking point or point flank was a very hazardous duty.

This particular day, I was walking flank point along with three other guys about 50 meters from our company. As we made our way through an open area surrounded by jungle and trees, I walked directly over a covered and well-disguised foxhole hidden from my view. As I continued to advance another 30 meters, suddenly, two gooks (Viet Cong), jumped up out of the "spider hole" firing their Russian-made AK-47's (the weapon of choice among communist insurgent fighters around the world) and started running across the field. The three Korean War and World War II-era Vets who were with me yelled at me to take cover. Diving to the ground, I rolled towards the enemy firing at them as they ran away. The Vets immediately ran after the two gooks. As I took up the chase, one of the Vets looked at me, pointed his finger in my face and said, "You stay here!" I uttered, "But." He pointed at me again and said, "No, you stay here. We got this." I figured they thought I might get in their way. It made sense; I was a rookie. It also crossed my mind the Vets were looking out for my best interests. Obviously, the Vets were trying to save my life. They returned in a hyper state of excitement. They were all covered in blood and were eager to share with us what had happened. After killing the two gooks, they had meticulously scalped both of them and had watched as their brains came out. The ritual mutilation was completed by cutting off the 82nd Airborne patches from their uniform and tossing them on their bodies.

The 82nd Airborne patch, which is worn shoulder height on the uniform sleeve, was the most distinctive worn in Vietnam. Whereas other unit patches hid their insignia with camouflage, the 82nd Airborne proudly and defiantly displayed their All-American red, white and blue colors. Upon finding the remains of those two dead gooks, there would be no doubt about which unit had done it. Additionally, according to the Vets, they mutilated the bodies because the Vietnamese believed they could not enter into Heaven with body parts missing. During my tour of duty in Vietnam, I never confirmed if this purported belief among the Vietnamese was fact or fiction. These incidents of mutilation were common among the Airborne and Special Forces. Other infantry men wore necklaces made of severed ears as a means of keeping track of their personal kills or body counts. Frankly, the sight of these mutilations didn't cause me much discomfort. But on the other hand, I also didn't feel the need to participate in this gruesome trophy collecting activity.

To my way of thinking at that time, there was no need to desecrate the body of a fallen enemy soldier with mutilation. I saw no need to dishonor a dead opponent. Besides, I would keep my body count tally by cutting notches on the butt of my M-16 rifle. My rifle was unmarked at this point in time. As I felt myself becoming more numb and indifferent at the gory spectacle of mutilating dead bodies, I asked myself a couple of questions. Was I already becoming truly hardened and callous? Or was this growing numbness and indifference a psychological defense mechanism to cope with the horror of war? But the day was still young and there were other things on my mind. Later that same afternoon, I would have my first confirmed kill.

That day, we encountered heavy small arms fire from about 50 NVA troops and several VC. By the time I arrived in Vietnam in 1968, the North Vietnamese Army (NVA) had essentially supplanted the home-grown resistance of the Viet Cong against the U.S.-backed South Vietnamese government. While the VC was a highly motivated guerrilla force, by contrast the NVA troops were a conventional army: well-trained, well-armed, and experienced battle-hardened troops. The NVA were now our primary adversaries on the battlefield. They were well hidden in the heavy vegetation and palm trees. Remember, we were in a vast jungle, but the area of this fire fight was confined to about two lengths of a football field and the depth of four. Suddenly from about 45 meters away, I spotted a gook with an AK-47 at the ready, running directly towards me! I do not remember being nervous. It was like shooting at a pop-up target exactly like the ones in our training. However, this pop-up target was armed and coming right at me! Reflexively, and without thought of self-preservation, all my previous training kicked into action. In that moment of truth, I found myself standing tall. It was time to lock and load and rock and roll.

During my Advanced Infantry Training at Fort Lewis, I was an expert marksman with an M-16 rifle in my battalion. So when I opened fire on the onrushing gook, I knew I had hit him several times from top to bottom. However, the VC kept attacking and firing wildly at me. As the rounds flew past me like angry bees, I popped in another magazine into my rifle and finally dropped him with another burst of fire to the head. To this day, I don't know how he missed me. After this firefight, I was told that the enemy often used opium prior to engaging in battle. Opium removed the natural hesitation to attack and numbed the pain when they got shot. We found traces of opium on his dead body; our suspicions were confirmed. Minutes later, we were on to our next firefight. From about a distance of 40 meters, two VC jumped up and started firing at us. I could see the enemy partially hidden by the vegetation; my first impulse was to throw a grenade. My athletic instincts kicked in at exactly the right time. The rocket arm I had become renown for while playing third base delivered the coup d'grace to the unfortunate VC. I hit him in the head and one arm flew up and then he fell to the ground. A second later there was an explosion. I heard someone in my squad say, "What the hell did you just do? Holy shit, do that again!" We were never quite sure if my powerful, accurate hurling of the grenade killed the VC on impact or was it from the ensuing explosion. A blast of adrenaline was shooting through my body and the rush was unbelievable. Either way, the guys in my squad thought it was hilarious. I was scared to death and these guys thought it was funny. They actually wanted me to throw another grenade just for the hell of it.

It was-- to coin a phrase-- a day to end all days. It certainly was a day to remember. There would be more days like this, but I would have been content to get on a plane and go home and consider myself a very lucky man. This would prove to be the first of several times that I stared into the face of death and escaped its cold, sinister eyes.

People always ask: "How does it feel to kill another human being?" In that moment of truth, I felt no remorse about killing another human being. Dehumanizing the enemy to the status of a pop-up target is what infantry training is all about. The sacred and deadly duty of every infantryman is to kill the enemy without hesitation before he kills you or your comrades. It's self-defense. That is how you survive in combat. I had no choice. It's either him or me. Unlike the movies, you don't look to the sky for forgiveness or fall to your knees and cry. I just killed a man; it was as simple as that.

It's my basic nature to internalize my feelings. Eventually, an "I don't care" attitude and operating without a lot of emotion became the norm for me. I became cold-hearted in some respects. That is just the way I had to be in order to survive. Unfortunately, that "coldness" stayed with me for many years after returning from Vietnam.

"No dumb bastard ever won a war by dying for his country. He won it by making the other poor, dumb bastard die for his country"---General George S. Patton

While a firefight is definitely a heart-pounding, adrenaline-pumping experience, the thought of an actual physical, life-and-death struggle with the enemy in hand-to-hand combat has always haunted me. I imagine the experience of going through lethal, intimate, close-quarter hand-to-hand combat would have had a huge residual impact on me. Even now, I am still plagued with nightmares about someone coming after me with murderous intentions.

Despite major technological changes such as the use of gunpowder, the machine gun in the Russo-Japanese War and trench warfare of World War I, hand-to-hand fighting methods such as bayonet remained common in modern military training, though the importance of formal training declined after 1918. During World War II, bayonet fighting was often not taught at all among the major combatants; by 1944 German rifles were even being produced without bayonet lugs.

Finally, the shooting settled down; although it seemed like hours and hours of shooting, throwing hand grenades, and dodging bullets, it was probably more like an hour and half. We then pulled back about 20 meters from the vegetation and set up a perimeter. An hour or so passed without any action, so we ate dinner. It was bizarre to be eating since we had just been in such close contact with the enemy. However, the captain must have felt the enemy would not leave the security of the jungle during this time. We were told the choppers were bringing in more ammo. The order was to "dig in and stay alert and wait on the perimeter lines."

Later that day the main topic of conversation were stories about my grenade throwing prowess. The whole platoon was thoroughly impressed with the velocity and accuracy of my grenade throwing ability. Whenever we needed a laugh, that grenade throwing incident was a ready source of material. After 27 days in the field, a couple of choppers arrived at the fire-base loaded with ammo, grenades, hot food and most importantly… MAIL! I'd been "in country" almost a month and at long last I was receiving my first mail. It was great! There were a couple of letters from my Mom and my sister and at least 20 or so letters from my girlfriend. Every G.I. enjoyed and appreciated a letter from home. Sadly, there wasn't much time to read our mail. But, it was just enough time to allow us to escape the harsh realities of war…at least for a few brief moments. Those letters had the effect of putting us in a comfort zone and reaffirmed that we were in Vietnam fighting for a reason.

The letters from home also reminded us how much we wanted to get back home to our family and loved ones. It might seem like a corny cliché to say this now, but those sentiments were genuine and heartfelt. Most of the letters from home asked the same questions: How was I doing? Did I need anything? What was Vietnam like? I can only imagine the anxiety and concern my family was feeling for me. Surely they were watching the TV news every night following the progress of the war and hoping to get some information about me. After reading my letters, I settled in with the others for dinner. When the Vets noticed I was eating spaghetti with red sauce, they couldn't resist having some fun at my expense. "Hey! What's that you're eating? Maybe I should toss my 82nd patch over that "brain" food you're eating!" The Vets were exploiting a great source of dark humor at my expense. Conjuring up those images of VC brain matter almost made me lose my appetite for the spaghetti.

But I was starving, so I ate the "brain" food anyway. In the distance, I could hear the echoing sounds of small-arms fire. After dinner we dug in for the night. We all dug foxholes three feet deep and five feet wide. But the night proved more frightening than the daytime. The stillness of the jungle night, along with the knowledge that the enemy was lurking highly intensified all of my senses. We knew for sure that the enemy was in the jungle not more than 20 meters from our perimeter. Off in the distance, I again heard a scuffle of some kind. It quickly passed.

What you can't see, but only hear what "goes bump in the night," is often much more frightening than the actual firefight. Needless to say, I didn't sleep much that night. Most of the evening, I stayed awake thinking. Out there just beyond our perimeter was someone who wanted to kill me. Someone who didn't even know me wanted me dead.

All was quiet on the jungle front. As the morning sun came up the following day, I was truly relieved that I had made it through the night. Two enemy soldiers had not been quite so lucky. Apparently, during the night two VC had infiltrated our perimeter. These two VC then proceeded to attack and engage three guys from our other platoon. That was the scuffle I had heard the previous night. It was the final exit from the battlefield for the two enemy intruders. The death of those two VC keeps your mind wondering about when it might be your time to engage in hand-to-hand combat whether you like it or not. Following the Captain's orders, a squad tossed the two dead VC in foxholes, poured gasoline on them, tossed in a lit match and burned them. The sight and smell of burning human flesh is something I hope I never have to smell again. The odor of burning human flesh is without a doubt the most horrendous stench I've ever experienced. The smell stays in your nostrils and haunts your memory for a long, long time. That stench stayed with me even after my return home.

That morning, we encountered more small arms fire. We were acting on intelligence reports that a huge number of both VC and NVA were in a nearby village. We were avoiding cutting through the dense jungle because it would have exposed us to enemy sniper fire or possibly tripping booby traps. The Captain surmised we were probably outnumbered. He ordered an air strike. The target for the air strike was the outskirts of the village. As the jets flew past to deliver their payload, it seemed like they were incredibly close. I could almost make out the names of the pilots on their helmets. They couldn't have been more than 50 to 60 meters above us! The roar of the jet engines rumbled and reverberated throughout my body. As the canisters of napalm exploded on the ground, the heat waves drew beads of sweat on my face and arms. Upon impact, the hair on my arms curled up. The heat emanating from the smoke and flaming inferno of the napalm strike erupted into an awe-inspiring sweeping spectacle of exhilarating beauty. Yes, I do indeed "love the smell of napalm in the morning."

Napalm Explosion dropped by F-4E Phantom Jet

A Walk Through the Valley of Death

CHAPTER SEVEN

The next few days we were occupied with Search and Destroy missions (S&D). We'd flush out VC or NVA from an area as well as go through villages suspected of harboring the enemy or their supplies. While rummaging through their huts (aka "hooches") usually made of straw, mud, twine, and cardboard from our c-rations cases, we'd often find hidden caches of weapons, medical supplies, or food. On this particular mission, we passed through the jungle into the village where we had called an air strike the previous day. All we found were traces of blood and a handful of destroyed weapons. There were no bodies. The enemy was pretty good at recovering and removing their dead comrades. I was glad I survived all the action from the previous days, and that this particular mission had come to an end. I was still alive. Unfortunately, there was news of another death on the home front.

ROBERT KENNEDY IS DEAD, VICTIM OF ASSASSIN; SUSPECT, ARAB IMMIGRANT; ARRAIGNED; JOHNSON APPOINTS PANEL ON VIOLENCE"---New York Times front page headline, June 6, 1968.

I was heartbroken when I heard the news that day, another Kennedy brother taken from us by an assassin's bullet. I was an admirer of JFK as well. It bothered me that RFK had been killed in my home state of California. In 1968, Bobby Kennedy had been running his campaign as an anti-war candidate. There was much speculation and hope at the time that a Robert Kennedy administration would have ended our involvement in Vietnam. I believe he would have ended the war and brought the troops home.

The next couple of months we continued search and destroy missions and night ambushes. We went through wet rice paddies where the water and mud was up to our thighs. There was also water buffalo crap to deal with as well. Water buffaloes were used to pull the plows to create furrows to plant rice. It was during a river crossing that I had my first encounter with leeches. A leech looks for any patch of skin on your body and attaches itself to it with bad intentions. Once attached, the leeches begin their tenacious sucking process. Faced with a leech stuck on my body, I had three options to rid myself of the blood sucking parasite.

Option One: Take out my handy dandy Zippo cigarette lighter and burn them off.

Option Two: Assuming I still had some mosquito repellent, I would pour the liquid over the leech and it would fall off in short order.

Option Three: This was the least desirable and most painful option.

If I simply let the blood sucker gorge itself at my expense, the leech would eventually get so fat and heavy it would simply fall off of its own bloated accord. Sometimes, Option Three was the only option because if the enemy was close by, burning or drowning the leech was not a viable choice. At times like those, the pain of a sucking leech stuck on my body was not a high priority.

We did not see much action in the northern part of South Vietnam. What we did get to see was Agent Orange falling from the sky.

Defoliation Program

At that time, we thought it was a good thing. It killed all the foliage and a few small insects. The Agent Orange fell all around us like rain falling from the sky. The military upper brass used it without hesitation or consideration of any unforeseen, long term side effects on the Vietnamese and U.S. Ground forces being exposed to the chemical. Unfortunately, for many G.I.'s who served in Vietnam, Agent Orange stayed in our systems long after our return to the States. The human devastation brought on by the deadly Agent Orange defoliate is incalculable. While I have experienced my own side effects from Agent Orange it unfortunately was fatal to several of my close friends. Even now, many others still suffer. Thank you Dow Chemical.

This particular day, I was in the vulnerable and exposed position of "point man" ahead of my squad on an open path. I see a circular ditch

The effects of Agent Orange

and I wanted to check it out. It appeared to be a bomb crater. It was about 9-10 feet in circumference and 8 feet deep. Leaning over to look into the ditch I grabbed a tree branch. As I leaned on the branch, it broke and I lost my balance. In that instant, I could see what was in there. I thought to myself, "Shit, I'm done." I was falling into the pit. Fortunately, my "slackman" grabbed me and pulled me back. I was able to grab the tree truck to push myself away. Within the eternity of those 5 seconds, I felt my life was about to end. But for the second time, I saw the face of death and avoided the fatal consequence. It was a booby trap! The "pungi pit" had been set up by the VC to impale some unfortunate G.I.'s on the razor sharp bamboo stakes positioned on the bottom of the pit.

These pungi pits were indeed very effective at catching an unsuspecting soldier off guard, especially at night. However, this particular pit had five

A Pungi Pit

Vietnamese civilians in it who were not so lucky. They were all dead. We concluded these villagers had been captured and interrogated by the VC and then thrown in the pungi pit. The bamboo stakes had pierced right through their bodies. The putrid odor emanating from the dead bodies stayed with me along with the gruesome visual. We were ordered to retrieve their bodies and go through their belongings and search for any valuable information. It was possible these civilians had information on the VC or NVA. The possibility existed they could have maps, dates or meeting times or even a location of weapons and supplies. But our subsequent search of these civilians resulted in no usable or valuable intelligence. Unfortunately, these poor, innocent Vietnamese civilians just happened to be in the wrong place at the wrong time. We were ordered to take their remains to the nearest local village to be identified by friends or relatives for proper burial. To be totally honest, nearly falling into the pungi pit was a traumatic enough experience for one day. I had nothing to do with searching the bodies or removing them from the pit. The other men from my platoon had to do it. There was no way I was sticking around! Meanwhile, I attended to more immediate concerns. I needed to pull myself together and stop trembling in my boots after another near brush with death.

We moved on. Two nights later we were told we were in a "hot zone." Being in a "hot zone" meant the enemy was extremely close. I set up claymore mines, and trip flares. For our perimeter, I set up my specialty; booby trapped trip grenades. Setting up a claymore mine was a relatively easy task. A claymore mine was like a mini-explosive. It was a small, thin suitcase like device filled with pellets or steel balls. The claymore mine had a slightly angular shape and the words "front toward enemy" written directly on the mine. The mine sat on scissor-like folding legs. It sat about six inches off the ground. An electrical wire connection was attached to the case. The wire could be as long as needed. At the other end was a remote control device called "the clacker." The clacker allowed a soldier to ignite the claymore mine from a safe distance. When the mine exploded, it shot out steel balls in a 60 degree arc from the device. These steel balls could inflict serious injuries and could be extremely lethal. Claymore mines were usually put closer to the perimeter in order to be used as an anti-infiltration device against the enemy.

Claymore Mine

Setting up a trip flare or my special grenade booby trap required skill and a steady hand. A trip flare was a canister-like device. It sat about 8 inches off the ground. When activated, the canister began to burn and it shot a flare up into the air. It was usually used as an anti-infiltration device. The bright flare illuminated the visual field. It permitted the soldiers within the perimeter to see an oncoming attack or encroachment of the perimeter. Trip flares were mounted farther out on a perimeter than a claymore mine.

Trip Flare

A particular skill was required when setting up a trip flare. That skill came into play when you attached the thin fishing line wire to the canister's ignition pin. You had to be careful when pulling out the pin just enough not to set off the flare. When someone "tripped" the wire, often brushed accidentally by the ankle or leg of an unsuspecting infiltrator crossing through the line, the pin was pulled and it triggered the flare. My special grenade booby traps were similar to the trip flares. I'd replace the trip flare with a grenade. The grenade was attached to a stake and then stuck into the ground. When it was tripped, the pin would be pulled and it would detonate. When setting this trap, I would pull the pin out as far as possible without it falling out so that it was very sensitive to movement.

Even though we were in a hot zone, the night passed without incident. The next morning, one of the team leaders went to disarm and retrieve the booby traps, unaware that I had set them. I had a rule that no one was allowed to disarm my flares or grenades but me. My traps were so sensitive that the slightest bit of movement would trigger an explosion. This is known as being "hair-triggered". When this guy, named Gardner, tried to dismantle my trip flare, it went off and gave him second degree burns. He would have incurred worse, possibly death, if he had tried to dismantle my grenade booby trap. The goof ended up receiving a Purple Heart for being injured in battle! Go figure!

Purple Heart Medal

CHAPTER EIGHT

In August 1968, after only four months "in country", I was promoted to squad leader. This put me in charge of 10 to 12 men. Being a squad leader was a huge responsibility. Now more than ever, I felt personally responsible for the lives of each man in my squad. I would get my orders from an officer at a command post meeting. He gave a general description of the mission and shared any pertinent information of what to do or where to go. Then, I was responsible for telling the men in my squad how to implement the directions I had just received. I appointed guys in the squad I knew could handle the tasks. It was the key to survival. These assigned tasks included telling them where we were going, who would set up the perimeter, appoint someone to walk point and back-up point. However, if a mission seemed particularly dangerous, I walked point.

There was disturbing news coming out of the Democratic Convention in the late summer of '68. We received the report from one of the men in our squad who had a portable radio. The report stated that there was large numbers of protesters demanding that the Democratic platform should include ending U.S. involvement in Vietnam. Tensions between the protesters and the Chicago police erupted into full-scale riots in the streets. Hundreds were beaten by the police and arrested. We heard that many of the protesters were Vietnam Veterans. These battle-hardened vets had faithfully carried out the U.S. government's military strategy of a "limited" war. They were angry and disillusioned that their policy was resulting in the senseless deaths of American troops.

Unlike World War II, there were no established front lines of battle in Vietnam. The Vietnam War, under General Westmoreland, had become a sectional, piecemeal war of "search and destroy" missions. For instance, we would get orders to take over a hill. We would go in, get control of that piece of territory for two to three weeks and then move out. When the VC knew we moved out, they would return to that area and occupy it again. It was a futile and vicious cycle. We were also fighting in a territory that we knew nothing about. The VC had total advantage fighting in the jungle. We were fighting the VC on their terms, in their own backyard! In my opinion, we didn't take full advantage of our military superiority and resources. Why, for example, didn't we continue the bombing campaign of Hanoi that President Nixon had started? I felt we could have bombed them into submission. All was fair in love and war, RIGHT? If our government wasn't willing to do that then, we should have pulled our men out of Vietnam and gone home! We were not going to win this war in the traditional military sense. There were feelings of resentment, disillusionment, and confusion that seeped into the mentality of the troops. There was no longer a good rationale for continuing this war and putting Americans in harm's way.

We were now doing some chopper assaults into "Hot Zones." The combat situation that unfolded during a chopper assault was mind-boggling. First, we were being dropped blindly into a hostile environment. Second, the enemy had a clear view of the helicopter and could see exactly where we were being dropped. Thankfully, on this particular chopper assault, the VC scattered when they saw us land.

View from the helicopter before landing.

Once again, we were out on a search and destroy mission. We followed off the trail as much as possible to avoid booby traps, ambushes and snipers. Finally, we saw a VC group with rifles and the chase was on! The VC fled into a semi-open area, but then disappeared. After searching high and low, we discover a tunnel. One of the men said, "I saw one of them go down into the tunnel." The captain asked for a volunteer to go down the tunnel. Without hesitation, I stepped up. I felt totally invincible. And why wouldn't I feel this way? This attitude of invincibility was common among my peers. It was born out of our commonly held belief that everything would be okay and that our goal was to get our job done and go home. We were living our 82nd Airborne Motto, "All the Way." At that time, I was young, single, no kids, no pressing responsibilities at home...

so why not? But the main reason I didn't hesitate to volunteer to go into the tunnel was because of what The Vets had given me, Knowledge!

Whenever there was an opportunity to learn some important piece of combat knowledge from The Vets presented itself, I took it. From those Vets, I learned much about the war, booby traps, walking point and most importantly for this particular volunteer mission...tunnels. I truly believe I owe my life to those Vets. Due to their guidance and patience, I gained a wealth of useful, vital, life-saving knowledge. Additionally, The Vets never gave me any attitude for my persistent and ongoing questions. They told me when I stopped asking questions they would be concerned about my well-being. It was those Vets that gave me my nickname "Smooth".

According to them, I had a certain "smooth" way of handling people to get what I wanted. This was especially the case when it came to women. My "smooth" talking abilities with the ladies had become evident to the Vets during our search and destroy missions. Oftentimes, we would come upon certain communities that were friendly. In these friendly communities, the locals would approach G.I.'s to sell their products as well as sexual favors. It was common knowledge that you could get the services of a prostitute for about five to six bucks. I would pick out the prettiest woman, I then negotiated with the "mama san" (any Vietnamese woman acting as a madam) and brought her price down to three bucks. Since I spoke some Vietnamese and they knew a little English, we managed to conduct business effectively. Because my smooth talking skills worked so well during these negotiations, my squad was a very happy and satisfied group of G.I.s!

Being called "Smooth" by The Vets meant a great deal to me. It meant that I had earned their respect. Just before The Vets left, they gave me a huge compliment. They told me that they would enter battle with me anywhere, anytime. The Vets said I was a true 82nd All-American. Those words of praise and confidence uttered by those guys made me feel extremely honored. There are no proper words to describe the respect I had for them. It was with that feeling of pride and invincibility that I entered the tunnel.

Entering a Tunnel

Inside a Tunnel

The entrance to the tunnel was about three feet in circumference. Armed only with a flashlight and a .45 caliber semi-automatic automatic pistol,

I entered the tunnel head first. Try and picture this dangerous scenario. Here I was, a six foot, 135 pound grunt trying to get down a dark, narrow tunnel. There was no way to know what to expect. About four feet down, there was a dirt floor. Moving forward from there required me to duck my head. The tunnel was three feet wide and at least three feet high. I cautiously crawled forward on one hand and two knees; sometime just on my knees. It was common for the VC to booby trap these tunnels with trip grenades, pungi sticks and venomous snakes. Like Indiana Jones, I HATE snakes! All my senses were on high alert keeping me fully aware of my surroundings. Suddenly, a noise! It sounded like someone was bouncing off the tunnel walls. I went from my knees to my belly. After crawling straight ahead about 20 meters, I hit a "T" intersection. Which way to go? Do I go left or right? The hot, thick humid air in the tunnel made my labored breathing even more difficult. My sweaty body was drenched with the gnawing sense of apprehension of not knowing what danger lay ahead of me. There were faint sounds coming from my right indicating the enemy was moving away from me. I decided to go to the right. I could tell there were about three to five VC ahead of me. The odor of human waste down in the tunnel was overpowering. At this point, my nose became my most valuable tool. That valuable tool told me to follow the unmistakable scent of VC shit. Due to their primitive battlefield conditions, the VC rarely, if ever, had the chance to bathe when they were in the field. Thus, the telltale odor made them easy to track in the tunnels. Inside the tunnel it was pitch black with my flashlight providing the only illumination. By now, my body was soaked in perspiration, palms slippery and moist, heart pounding against my fatigues like a jack hammer. I was filled with an overwhelming sense of anxiety. Contact with the enemy was possible at any moment. Inside this tunnel, contact with enemy meant either death for him...or me. I was in the tunnel alone with no back-up. I was scared, very scared. Damn! I hit another "T" in the tunnel. The subsequent twenty foot crawl seemed like a hundred yards. The tunnel was a dark maze. Hoping to find my way out, I made twists and turns. There was still no encounter with the VC at that point. In one sense, I was glad. However, in another sense, I wondered if the enemy was waiting for me just around the next turn. I would stop and listen. There! A noise! It was something in front of me. I pointed the flashlight in that direction and it revealed it was only a rat. A rat that appeared to be more scared than I was. Obviously a fellow "Tunnel Rat!"

I continued to crawl on my knees for another thirty feet, I saw an open area. No enemy in sight. But there was a cache of weapons and miscellaneous hospital supplies. The area appeared like the rest of the tunnel. It was spotless, as if someone had just swept. There was no VC in sight. I figured they had crawled out of another tunnel exit and fled the area. It was common for there to be two to three different exits from the tunnels. A tremendous feeling of relief washed over me when I found my way out of that tunnel; I had not made contact with the enemy. Mentally and emotionally I was thoroughly exhausted. Crawling around in that dark, foul-smelling, dangerous tunnel was

A cache of weapons and supplies.

a very petrifying experience. But it was one helluva RUSH! After we secured that tunnel, a five man team went down to clear out all the weapons and supplies.

Following that, we threw grenades in the tunnel openings to seal them off and it was our aim to make them unusable. Afterwards, we moved on to another site. To a casual observer, volunteering to be a "tunnel rat" would seem like taking a deadly, unnecessary risk. During my tour of duty in Nam, my motto was to keep everyone alive and make it back home. Therefore, I felt it was vital for me to learn and know how to do everything. These combat skills would increase our mutual chances of survival.

CHAPTER NINE

The next day we settled into a fire base. This area was designed as a safe haven located within a defense perimeter. On this day, the winds picked up. Though it was still hot, there was something in the air that gave me an uneasy sense of instability. But compared to what I had previously experienced, the next two weeks at the fire base were like Heaven.

A Fire Base

Me in front of a bunker, 1968

We actually got to sleep inside a bunker. We wrote letters home and even tossed a football around. Occasionally, the supply choppers in addition to delivering ammo and mail brought us hot food. The choppers played a significant role during my tour of duty in Vietnam. The giant Chinook (CH-47) helicopter transported large numbers of troops from one area to another. The Huey (UH-1) was a multipurpose helicopter used for transporting troops and to medivac wounded soldiers. The Cobra (AH-1G Below) was an attack helicopter also known as "gunship" armed with rockets and machine guns.

Besides the bunkers and hot food, we also had the protection of nearby artillery. Occasionally, a tank would make an appearance in our area. The listening post about 50 meters outside the perimeter manned by a Radio Transmitter Operator (RTO) added to our sense of security. In contrast to

Chinook

AH-1G

sleeping in a muddy jungle or wet rice paddy, the fire base bunker was a welcome, if only temporary, safe haven. It was during this time at the fire base that I decided to write to my girlfriend back home and tell her that we could no longer be engaged. By this time, I had already seen so much death and experienced so many dangerous situations. Although I felt invincible, there was the possibility that I could be killed any time. Also, knowledge of my sexual escapades in Vietnam would most certainly put a damper on my relationship. I received a letter from my fiancé indicating she understood that it was an unfair relationship. However, despite the drastic shift in our relationship, she continued to write to me.

Well, all good things come to an end. Our heavenly, leisurely days at the fire base concluded abruptly. I still had these uneasy, premonition-like feelings of some ominous presence in the atmosphere. I was right. Our fire base was hit hard by a severe typhoon. I was asleep on the bottom of a three tiered bunker that was protected by heavy sand bags and airfield runway steel plates. These forged steel plates weighed 150 pounds each. The twenty foot long plates were placed between and secured by the heavy sand bags creating the three tiered bunker. When the 120 mph typhoon winds ripped the top section off the bunker, the three tiered bunker quickly became two-tiered. The two-tiered bunker became one-tier when the second section collapsed. Suddenly, upon waking, I was faced with the horrifying possibility of 1500 pounds of sand and steel crashing down on me. There I was helpless and trapped with all that crushing weight of sand and steel only three inches angled from the top of my head to my feet. Terror-stricken, I screamed for help. At any moment that 1500 pounds of sand and steel was surely going to collapse; crushing me to death. Everything fell silent. Everything was still. Just then, the opportunity to slide out of my "grave" presented itself. I slid out and looked back. I was surprised and relieved when I saw my M-16 lying at such an angle that the rifle sight caught the bunker tier, preventing it from crushing my head. I had narrowly escaped death for the third time. I loved my rifle. There were so many ways my M-16 had saved my life.

U.S. Military issue M-16 automatic rifle

Steel plated landing strip

Three-tiered bunker

Two days later we were out on patrol. It was hot. The 108° temps were hotter than usual. We humped (walked) about 20 klicks (a "klick", abbreviation for kilometer, 2/3 of a mile). I walked point during this outing and I was totally exhausted. As we approached base we went through a dense area of vegetation. My back-up, or slack man, who is the first person behind the point man, said, "Hey Smooth, you got a tripwire wrapped around your ankle." In bone-weary disbelief, I stopped, looked down and thought, "Oh shit! I really don't have time for this!" Sure enough, it was a booby trap. After walking point for 4 months, an enemy booby trap was wrapped around my ankle and had me in its deadly grip. My emotions bounced back and forth like a pin ball machine lit up on speed. I was extremely angry, afraid and just plain tired of it all. Finally, after what seemed like an eternity, in a weary instant of fatalism and resignation, I looked down and was shocked and amazed that the pin was not in the grenade, but somehow was still attached to the grenade. I assumed that it was a dud I reached down and grabbed the grenade from my ankle and flung it about 20 meters away. My men were amazed that I lifted the grenade from my ankle without it going off. Seconds later, it exploded! The grenade on my ankle incident was yet another instance where I had narrowly escaped death for the fourth time. It seemed like Lady Luck had smiled on me again. We headed back to the fire base. By the time we got back to the fire base they were out of hot food. That night, after a long day on patrol in the sweltering heat and stifling humidity, even the canned rations tasted good. It was good to be alive.

Out on patrol the next day we were back in a jungle maze of dense vegetation consisting of thick plants, trees, and bushes. That night the temperature cooled down to about 85°. It was time to get some sleep. My humble and primitive sleeping arrangements consisted of lying on the dirt, using my helmet as my pillow and my poncho liner as a blanket. Taking care of "when nature calls" was even more primitive. If I couldn't find a bush, I just peed out in the open. Taking a dump was a little more complicated. First, you had to find a spot and then dig a hole. If toilet paper was not available, a copy of "Stars and Stripes" newspaper came in handy, or as a last desperate measure, a handful of big leaves did the trick. A moon-lit sky and what seemed like a million stars provided the evening light in the jungle. The stars hanging in the sky seemed so close. So close I could just reach out and touch them. I spent a good portion of the night fighting with mosquitoes that were trying to suck my blood. Weeks went by without much action. During this time, I wrote letters to my family. Lacking any proper writing stationery, we improvised with whatever was available. I wrote my letters on empty C-ration boxes. By unfolding the C-ration cardboard boxes, I created my own letter writing stock. My letters to family and friends were always filled with confident reassurances that the war was almost over. "Don't worry about me. I got everything under control." I would write. And then jokingly add, "I'm just vacationing here in Flower Land." In their return letters, I felt they enjoyed my positive attitude. Soldiers in the midst of combat traditionally don't want to worry the folks back home more than absolutely necessary. When we were through writing, we folded our cardboard paper into envelope size. A mail clerk would take our letters, put them in envelopes

Reflection

with stamps and get them mailed back to the States. My family thought it was great receiving a letter written on cardboard.

Also during this lull from the action, it gave me time to think and reflect. I would sometimes get into a deep philosophical thought of how I was really two different people. Here in Vietnam I was the combat warrior in the jungle and the other was a kid from East San Jose. I remember taking a picture in Vietnam where I'm holding a piece of a mirror and you can vaguely see my reflection. This picture was my depiction of the opposing personalities.

For a couple of days, we stopped receiving C-rations. We were down to cheese, crackers and a couple of chocolate tropical bars. The chocolate was similar to a thick Hershey bar but without the almonds. The food shortage situation got so bad we had to forage for our own food. Some of our guys from the South decided they would go fishing to satisfy their hunger pangs. Trip wire was used for the fishing line. A man-made hook tied to the line and a branch for a pole completed the improvised and ingenious fishing gear. This was real Huckleberry Finn stuff. They sat along a narrow river bank and cast their lines into the water. However, they were not very successful. Finally, I said, "Hey guys, I've never fished before but I bet I can catch a fish." They looked at me and

Fishing in Vietnam

laughed. "Sure Smooth, Whatever you say." "Okay, I bet you guys five bucks I can bring in a catch." They couldn't resist. "You're on!" They had taken the bait. I chuckled to myself as I cautiously removed a grenade from my ammo belt, pulled the pin, tossed it in the water and yelled, "Fire in the hole!" The grenade exploded in the river like a depth charge dropped at sea to blow up enemy submarines. To my surprise, several edible fish floated to the surface. You don't need a fishing pole when you have a grenade. I collected my five bucks and went on my way.

Military Money

Vietnamese Coins

A Walk Through the Valley of Death

CHAPTER TEN

Several days later our company commander received intelligence of VC in the immediate area. We saddled up and once again we were on patrol. Within a half hour, I saw a VC jump up from a covered foxhole. Mickey McKay, the son of a well-to-do family from Cape Code, Massachusetts, was walking flank point. Mickey was also still a "cherry." I yelled out to him: "Mickey! Take cover!" Like a scared child, Mickey scrambled to take cover behind the first thing he saw. He dashed behind a tree. No doubt had he not obeyed my order he would have walked right into a bullet. The AK-47 rounds whizzed past Mickey like a laser beam. Afterwards, shaken and still with a look of sheer terror in his eyes, Mickey gave me a grateful nod of appreciation, "Thanks Smooth. I owe you one." Mickey waited patiently behind the tree for us to come to him. There was several other VC hiding close-by. They abruptly broke cover and ran. We were in hot pursuit of them and then we opened fire. We killed two and wounded one. In the distance, we could hear the others noisily fleeing ahead. Then just as quickly, they disappeared. A grunt from another platoon said, "I saw the VC go down a tunnel."

By this time, I had already been down about four tunnels. It was mostly in reconnaissance without enemy contact. I was relieved that I only found caches of weapons and medical supplies. However, in the tunnels nothing was certain until the search was over. Because of my experience, the captain came to me and asked, "Smooth, we need a volunteer to go into the tunnel, how 'bout it?" I thought for a second and said, "I don't have a good feeling about this one Captain; maybe we should smoke them out or grenade them." Our company didn't have dogs trained to go down into tunnels, so that was not an option. I suggested that we drop some colored smoke into the tunnel and wait. There was no smoke exiting and my suspicions were confirmed. There was no other exit. Nevertheless, the Captain asked me again, but this time tempting me with a "prize." "Go down into that tunnel Smooth and I guarantee you a Silver Star with V." Getting a medal was the furthest thing from my mind at that moment. My gut feeling and instinct told me it was definitely not safe down there. The Captain knew I was the invincible crazy kid who would do just about anything. Well, I was crazy…but not stupid! My answer was final. "I won't go." Sergeant Billy Smith volunteered for the assignment. Billy was an honest, church-going Catholic; an all-around good person. At 5' 9", 135 lbs, and blessed with classic Anglo-Saxon good looks, Bill was the perfect marriage candidate for any guy's kid sister. We were good friends and had the utmost respect for each other. That morning at church services, as we sat on C-Ration cases stacked on each other about three feet high. We listened to the Chaplin's sermon. I admired Billy's obvious devotion to God and his faith.

He came to me for advice about going down a tunnel. I didn't want him to do it and desperately tried to talk him out of it. "But Smooth, I feel it's my duty," he insisted. "If you have to do it, be very careful," I cautioned. "Don't turn your flashlight on until you hit the 'T' intersection." If he followed that procedure, Billy would still have cover from both sides of the wall as well as minimal visibility from the entrance. Since his flashlight was going to provide very little illumination and visibility, Billy would have to rely heavily on his nose and ears.

The captain ordered a troop to tie a rope around Billy's waist and then lowered him down into the tunnel. As they lowered Billy down, I thought back to when I went into a tunnel. "I'd never let them tie a rope around my waist," The whole idea of a rope implied an ominous purpose. For that reason, I always refused the rope. He went down with a flashlight and .45 caliber semi-automatic pistol in hand. He was only down there about 7 or 8 minutes when all Hell broke loose. Then I heard the unmistakable sputtering and rattling burst of a Russian-made AK-47 automatic assault rifle. There was just one short blast. I didn't hear Billy's .45 return fire. As I had feared, my premonition about Billy going down the tunnel came horribly true. The enemy had waited patiently for him and killed him. We furiously pulled on the rope and his limp, life-less body came up out of the tunnel. Billy had been shot dead. Shortly afterward, we removed from his body, his dog tags, pictures and any pertinent paperwork. Then he was carefully placed in a body bag. A medivac chopper was called to retrieve Billy and he was taken away. When he arrived home, Billy's grieving family received him in a flag-covered coffin; to me, his death should never have happened.

The captain then suggested we grenade the tunnel and smoke it. Filled with a toxic mixture of survivor's guilt and anger, I sarcastically asked myself, "Where have I heard that suggestion before?" Billy had been sent on an unnecessary, pointless suicide mission. The two VC in the tunnel were eventually killed by simply tossing in a grenade. The devastating effects of a concussion blast from an exploding grenade were horrific. These effects were magnified when the grenade exploded in a tunnel with no exit. A concussion force of that magnitude can blow out ear drums, liquefy lungs and in the case of the two unlucky VC in the tunnel…be deadly.

After Billy was needlessly sacrificed in the tunnel, I was extremely pissed off at the captain. What had compelled him to send Billy down into the tunnel when there was clear evidence it was not safe? I was overwhelmed with sadness and regret at losing my friend. That night, for the first time in a long time, I wept. That night, for the first time in my life, I questioned my belief in God. Why had God taken Billy? Billy had believed so strongly that God would protect him. I began to relive the whole scenario in my doubt-filled and guilt-ridden mind. "I should have gone down into the tunnel, not Billy," I kept thinking to myself. "Things would have ended differently. I had more experience." Maybe I should have insisted that no one should have gone down the tunnel. Maybe I should have been more forceful with Billy and talked him out of it. "I should have gone down into the tunnel, not Billy."

The unrelenting feelings of regret, doubt, guilt and anger revolving around that incident still torments my soul to this day. Now I know that what I have been feeling all these years is survivor's guilt. The mental anguish in remembering all the times I should have, could have, died and didn't. The terrible image of Billy's lifeless, bloody, bullet-riddled body being pulled out of that tunnel has been the source of many recurring, inescapable nightmares and countless sleepless, drunken nights. Beer and tequila help, but not much. Often times, drinking makes it worse.

CHAPTER ELEVEN

It was another day walking point. Do I forget Billy? Not now, not ever! I'm walking point with sixty pounds of gear on my back and I'm hacking my way through "wait-a-minute" vines.

These are vines with sharp thorns similar to those on a rose bush. If you walked passed these vines with your sleeves rolled up and accidentally rubbed up against one, it was like being attacked and clawed by a jungle cat. With a sense of vigilant anticipation we continued to move forward. I kept expecting that at any moment, a sniper hiding in a bush, in a hole, or up in a tree, was waiting to kill me. This type of constant and intense anticipation of being killed or

Wait-a-minute Vines

maimed wears on you psychologically. It plays games with your head.

We humped another 20 or so klicks through creek beds, rocky hillsides-- which are more like cliffs-- and heavy vegetation. Our captain urged us to pick up the pace. Apparently we needed to be somewhere before dusk. Upon reaching our destination, I passed out from fatigue and dehydration. In a damp, cold sweat, I was in a hazy mental fog. To make matters worse, we came under an enemy mortar attack. Mortars were the scariest weapon. You could hear the resounding, familiar "thump" as they were shot off so you knew they were coming. However, during a mortar attack, there was no place to hide. This was especially true in the jungle where there were no bunkers. To this day, I can still make out the distinctive "thump" of a mortar shell being dropped into a mortar tube. I'm aware of the captain's presence and hear, "Who's down?" Somewhere in the fog another voice, the sergeant, says, *"It's Smooth."* "Shit," barked the captain, "We need to medevac him out now!" The captain ordered his RTO to call in a medical chopper for a "dust off." But the chopper pilot, gripped with fear, refused to land while under mortar fire. I heard the captain instruct the pilot in no uncertain terms, "I need you in here NOW!" The urgency in his voice as he insisted I be picked up revealed his great respect for me and that felt good. Fortunately, the mortar shelling subsided long enough for the chopper to come in and pick me up. Although I was having my personal belief issues about God, at that moment I couldn't help wonder if God had stopped the shelling so I could be picked up. Or more likely, it was just dumb luck. Either way, I was on my way to a Mobile Army Surgical Hospital (MASH) unit.

Soon after we flew away, off in the distance I could make out the sound of the mortars resuming fire. I was taken to a MASH unit. It was just like the one on the long-running TV series M*A*S*H. The hospital was my safe haven; a sanctuary from the madness beyond its walls. Being in the hospital was like being back in civilization. There was hot food from a real kitchen. I encountered people who were genuinely concerned about my safety and well-being. Best of all, was the pleasure of seeing women who looked like they could have been imported from my hometown.

A Walk Through the Valley of Death

CHAPTER TWELVE

After my fourth day in the hospital, a chaplain, Major Rogers, asked if he could talk to me. "Sure," I said. He asked me about my religious beliefs. I said "You've caught me at a bad time chaplain," I explained. He said, "Why?" I said, "Well, for a few reasons, first and foremost God and I are not on speaking terms right now." I continued, "I do not want to believe God let me down and/or Billy, but Billy is dead and I'm not ok with that, no matter what the circumstance. If God is our Savior, what does that mean? I've lost my faith. Were my expectations too high about God and his guidance or sphere of protection? Is this St. Christopher medal I have worn daily through this ongoing melee been a hoax? Too many questions, not enough answers." Chaplain Rogers inquired, "What type of missions have you been on?" "I've been walking point for seven months, gone down into too many tunnels, been involved in my share of ambushes plus several chopper assaults. "During all that time," he asked, "Have you ever been injured or wounded?" No, aside from a bad case of jungle rot on my ankle," I replied, "I've been pretty lucky."

That case of jungle *rot* on my ankle did have one very bad, long-lasting side effect. I was scheduled for five days of Rest and Relaxation (R&R), but due to the infection on my ankle I wasn't able to go to Australia. The chaplain was curious to learn more about the infection on my ankle. "The medic who treated my ankle said he wouldn't allow me to go to Australia on R&R because maggots were coming out of my ankle. He said the infection might spread and it was too risky. So, I didn't go." There are few sights as unforgettable as watching maggots emerging from your body. It's not something you think will ever happen to you until you have died. It's an eerie experience. I had always associated hordes of maggots with feeding on a dead body. Turns out the disgusting little critters are beneficial. Maggots actually eat infections and aid in promoting the healing process. Despite the maggot's medical usefulness as infection fighters, that incident was responsible for me eliminating rice from my diet. I avoid even looking at rice because of its resemblance to maggots. Spanish rice is especially offensive because of the spices and seasonings that are used to make it look exactly like my infected ankle. Now, Spanish rice grains are maggots crawling around nibbling away at my pink flesh. I could tell from the wincing, frowning expressions on the chaplain's face that he had heard more information than he had bargained. What can I say? He asked. Days later the chaplain approached me with an offer.

"I need a driver to take me out into the field when I do services. Are you interested?" I could hardly believe what I was hearing. "Are you pulling my leg chaplain?" I asked with a smile. "No," he laughed. Driving a chaplain around in a jeep sounded pretty good to me. Confident I was the perfect guy for this job; I accepted the offer, even though I was currently agnostic. I immediately wrote home and told my family the good news. No more combat missions for this grunt. My remaining tour of duty in Flower Land would be spent driving a chaplain around in a jeep. I might be reaching, but did this happen because of my lack of belief in God and now a servant for God is asking me to be his driver?

I was told later that when my Mother heard the news of my new assignment, she exclaimed with relief and joy, "Finally, my boy will be out of harm's way!" Their happiness at my good fortune was obvious in their return letters to me. It didn't seem possible that it had only been two weeks since I'd been brought to the MASH unit. And now here I was reborn as a jeep driver for a chaplain. I could not believe my good fortune and this was no ordinary jeep either, it had a .30 caliber machine gun mounted on the back! From the looks of it, I figured it could shoot for miles. Naturally, the temptation to fire off a few rounds was too great. I had to fire that .30 caliber machine gun.

M-151 Mutt

Upon opening fire with the .30 caliber it seemed like the jeep was going to pop up right off the ground. Luckily it stayed put. Every round I fired sent waves of sheer exhilaration coursing throughout my body. Could it be true? I was actually enjoying being in Nam.

A Walk Through the Valley of Death

CHAPTER THIRTEEN

"All good things must come to an end"---*Geoffrey Chaucer*

"Missed it by that much"----*Maxwell Smart*

My new assignment as a jeep driver for Chaplain Rogers was like a dream. A Hollywood screenwriter couldn't have written a better ending to my story in Vietnam. I had become comfortable and complacent during my brief time as a jeep driver. I could easily do this for my remaining five months. But things were about to change dramatically. I was abruptly brought out of my happy sleep walking state by Captain Cooper, an Army doctor from the MASH unit. "Are you the fella they call Smooth?" "Yes sir, I am," I answered. With a grim look on his face, he said, "There's a young man from your unit who wants to talk to you."

"Thanks for finding me sir." I entered the hospital tent. It was Jordan. I recognized him as soon as I walked in. A million images flashed through my mind. There was Jordan, the "cherry", who had only been "in country" a month. During one blazing hot day, Jordan, the southern boy with the lily-white skin, received second degree burns when he ignored our advice to keep his shirt on at all times while outside. The poor kid didn't get any sympathy from the squad as he was the target of merciless teasing and the butt of our jokes. I could hear the guys telling him, "You should have listened to us. Next time, leave your shirt on, Cherry boy!" But, this was a very different Jordan that I saw laying on that hospital bed.

There was no avoiding the seriousness of Jordan's situation. My eyes were riveted on the plastic bag containing his intestines hanging from the side of his bed. It was an unbelievably gruesome sight. He was on a respirator, but could still speak. I leaned over and whispered to him, "Jordan, what the Hell happened to you?" "I was walking point Smooth," he answered weakly. "And I got hit by a sniper." I couldn't hold back my rage, "Who the fuck let you walk point?" When I was in charge of the squad, it was strict policy to never let a new man walk point. I asked who had given him the order to walk point on that mission. Jordan's barely audible answer made my blood boil. "It was the new 2nd Lieutenant," he murmured. Those of us in the field did not have any respect for the "90-Day Wonder," 2nd Lieutenants fresh out of their ROTC college program. As soon as these guys got their little gold bar on their shoulder, they were sent to Nam to command a platoon of men in hostile territory with little or no real combat experience. Well, maybe I'm being a little too cynical by painting all 2nd "looies" as helpless dummies. I had met some 2nd Lieutenants who were smart enough to utilize the experience of the more battle-tested vets (usually a Non-Commissioned Officer or NCO) to get the mission done right without jeopardizing lives.

I was overcome with a sense of utter helplessness sitting next to Jordan. Holding his hand, I asked Jordan, "Is there anything I can do for you?"

His answer brought me to a fork in the road in my journey through Nam. Jordan was asking me to take a path I was hoping to avoid and was steering away from as a driver for the chaplain. "Smooth," he said softly, "Can you go back and KILL all those fucking gooks for me?" A tear ran down Jordan's face. In the few minutes I had spent with him in that hospital tent, everything had changed. I was transitioning from jeep driver back to combat soldier in order to honor a dying buddy's last request. Without hesitation, I said, "You bet your ass I'll go back."

After packing my gear and my M-16, Captain Rogers agreed to drive me to the chopper pad. On the way, the chaplain asked me very calmly, "Do you really know what you're doing?" "Yes," I replied resolutely. "I'm going back to even the score." Chaplain Rogers shook my hand, blessed me and wished me Godspeed even though he had reservations about me going out to kill another human being for the sake of revenge. Since I last saw Jordan in that hospital bed, I have had nightmares and recurring feelings of remorse. I have also been plagued with second-guessing myself regarding certain battle field tactic decisions I made. Like with Billy, I felt a strong sense of responsibility to keep my men safe. I should have been more forceful with Billy and not let him go down the tunnel. Had I been on that mission with Jordan, I would not have let him walk point. That's the duty of a squad leader. If I had gone down the tunnel, would I have been killed? If I was walking point that day, would I have been shot by the sniper? These are the unanswerable questions that stay with me to this day.

A Walk Through the Valley of Death

CHAPTER FOURTEEN

I was back in the "bush" again; "a grunt on the hunt." I returned to my position as squad leader and spent some time relaxing with my men and getting reacquainted. I never brought up Jordan's name. He died shortly after I left. Even though I was back with my squad, things were different. There was noticeable tension in the jungle air. We now had another 2nd Lieutenant named Blake. He was a prototypical WASP from a well-to-do family. He apparently was not well-off or well-connected enough to keep him out of the war. The lieutenant and I had a couple of run-ins. First, he tried to tell me which trail to take on the way back to base after a Search and Destroy mission. He foolishly wanted us to walk in the open trail where there had been snipers and anti-personal mines. I refused. The platoon ignored him and followed my lead.

When we got back to base, he reminded me about the chain of command. I told him in the least condescending tone of voice I could muster, "I completely understand your position, sir." That obviously wasn't good enough. He snapped back, "You disobeyed a direct order. You got shit burning detail." Shit burning detail was a small price to pay for nobody getting injured on the mission because of some bonehead decision by a green 2nd Lieutenant. Burning shit consisted of taking out all of the 50 gallon cut barrels from the outhouses and pouring gas in them and throwing a match into the cans. After all the shit has burned away, the barrels are returned to the outhouses. Yeah, it's a stinking detail. But the odor of burning shit isn't nearly as bad as the smell of burning human corpses.

My next "difference of opinion" with Lieutenant Blake was at night. We were in a hot zone and we had humped all day. I was sure the VC knew where we were set up for the night. The squad ate C-rations together and relaxed before guard duty. I went around to my guys to set up watch. Then I noticed two of my men were missing. "Sergeant," I asked urgently, "What happened to the two other men from my squad?" One of my guys said, "The Lieutenant sent them out on a listening post about 75 meters outside the perimeter." "That's insane," I replied angrily. Not only were my two men outside the perimeter 75 meters away from the rest of the company, they were out there without a radio! I immediately went to see the Lieutenant in private to discuss the matter with him. "Lieutenant, we need to bring my two men back within the perimeter, they have no way of communicating with us if they make contact" I also argued, "If they get attacked tonight, there is no way we can get help to them in time!" He flatly refused my suggestion. Once again he reminded me of the chain of command and that he was running the platoon and not me. I once again reassured him that I understood who was in command, but that we might be making contact the next day and there would be shooting from both sides of the fence. "I'm not sure," I concluded quietly, "that we can protect you." "Are you threatening me?" he asked. "You son-of-bitch," I replied, "If you don't get my men back right now, it won't just be a threat!" As his eyes glazed over and he started to tremble, he said quietly, "the matter will be taken care of." True to his word, the "matter" was taken care of with my men. I was, as were my two men, pleased to return to the perimeter. The cherry Lieutenant and I never had any other disagreements. I can't vouch for his mental stability when he left Nam, but I can honestly report that the Lieutenant made it home in one piece and had a better respect for the men in his command. However, this was not always the case. Early on, a 1st Lieutenant wanted to make Captain at the expense of the men in his command. He would make ill advised decisions in hostile environments that would put us in harm's way. Fortunately no one from our platoon was injured, but that was just pure luck. Our platoon made a request to the company commander to transfer the 1st Lieutenant to another unit. He was transferred to another company and three weeks later we read in the Stars and Stripes that the 1st Lieutenant had been killed in action; that's all I have to say about that.

"During its long withdrawal from South Vietnam, the U.S. military experienced a serious crisis in morale. Chronic indiscipline, illegal drug use, and racial militancy all contributed to trouble within the ranks. But most chilling of all was the advent of a new phenomenon: large numbers of young enlisted men turning their weapons on their superiors. The practice was known as "fragging", a reference to the fragmentation hand grenades often used in these assaults. Between 1968 and 1973, dozens of Americans and Vietnamese were murdered in fragging incidents, but only a handful of their killers were ever brought to justice"----George Lepre

I can honestly report, that during my tour of duty in Vietnam, there was no racial incidents or drug abuse within our company, nor were any superiors killed or injured.

A Walk Through the Valley of Death

CHAPTER FIFTEEN

Richard Nixon was elected president of the United States in November of 1968. The president's platform was getting us out of Nam "with honor." Not only did he NOT get us out of Vietnam "with honor," he actually increased the troop levels to a new high of 543,000. Nixon also violated the neutrality of Cambodia and Laos by authorizing illegal "incursions" into both of those neighbors of Vietnam. In the vernacular of the grunts, it was referred to as "going over the fence." In addition, during Nixon's presidency, the United States also suffered the highest number of American soldiers killed in action. Out of the 58,000 killed in action over 35,000 men were killed while Henry Kissinger was attending the Paris Peace Talks with North Vietnam. Nixon's aide, Anna Chennault was sent to the South Vietnam Embassy to inform the South Vietnamese government not to negotiate with North Vietnam until Nixon was elected president, because Nixon could get them a better deal. This also opened the door for Nixon to get elected, because negotiation stopped and the War would continue. Nixon's platform was to stop the war. Nixon after elected started the Peace talks again in early January of 1969. The war continued under Nixon until an agreement was signed in 1973, but the last of our troops did not get out until 1975. His negotiations failed as the North Vietnamese overran South Vietnam and currently control all of Vietnam, thus he never made a better deal for the people of South Vietnam. So the question remained: would Johnson have ended the war in 1968 shortly after he stopped bombing Hanoi if Nixon did not interfere? As my Native American grandmother would say, "I don't' care for politicians who speak with forked tongues." Grandma was born in the late 1800's and went through the era of her people losing an enormous amount of property in the West. Although the government tried to reciprocate with a government issued check to grandma for her losses, grandma told them to shove the check up their political ass.

Our battalion was moved down to the Cu Chi province located close to Saigon. The battalion commander and General Westmoreland were still concerned about underground movement after the Tet Offensive in January 1968. We were set up at a field fire base, which had artillery and a tank. After a couple of days at the fire base, things were pretty quiet. So, we were given two days leave in the capital city of Saigon. With a few drinks in me, I decided to get a massage. Some of my buddies said they had a massage and it was the best thing ever. At the massage parlor, some very attractive Asian women caught my eye. "I'd like to get a massage," I said. Two beautiful women who were not wearing much clothing escorted me to a room. It was a very pleasant setting complete with American music in the background. The intoxicating aroma of incense permeated the room. The two ladies promptly helped me disrobe. Nevertheless, I kept my .45 caliber pistol close by. Whenever I was in town, I always exercised great caution. These two women were amazing. One walked on my back using her bare, agile toes like soft fingers massaging my every muscle until all the muscles melted like warm butter. The other woman worked my toes, feet and legs. She was extremely gentle. Her fingers were soft and supple. Her magical hands hit every sore muscle with expertise. It was a great way to release all my built-up tension.

After leaving the massage parlor, I went to a lounge. There, I met a very nice Vietnamese woman. She was attractive, witty and charming. She had learned to speak English while attending a private Catholic school. As the evening wore on, we talked, danced and had a few cocktails. Her invitation to stay with her for a couple of days caught me off guard. Was this my smooth talking or were we just two very lonely people? She lived in a modest, one room apartment. We truly enjoyed our brief time together. No talk of war or politics. It was a good, mutually gratifying relationship. When it came time to leave, I asked: "Do you need any money?" "No," she said. "But I do need some 'D' cell batteries for my radio. And," she added with a smile, "I like chocolate." I took care of both of her requests. It was the least I could do for the great time we had together. This was an awesome two days away from everyone and everything. I went back to base the next day.

A Walk Through the Valley of Death

CHAPTER SIXTEEN

At daybreak we moved out without any confrontations with the enemy. A week later, in the southern part of Tanphu Trung, Cu Chi province, we saddled up once again for a night ambush. Our new company commander, Captain Sam Foy, who was with Special Forces, thought it would be a good idea to go out on ambush when it was completely dark. This was contrary to normal ambush procedures which dictated going out at dusk and reaching the ambush destination when it got dark. Nevertheless, we reluctantly packed up and moved out to our assignments. I was walking point on the dikes of a rice paddy. A rice paddy dike was a rectangular shaped structure. One individual section of a rice paddy dike was 2 feet wide, 2 ½ feet high and about 150 feet long.

It seemed darker than usual that night. I was very scared. Normally during these types of operations, I would be a little anxious, but never scared. As we continued to our ambush site, I noticed how damp I was. By the time we finally arrived, I was thoroughly soaked with perspiration. I can't emphasize often enough how frightening it was to be in complete darkness. Not knowing what is out there waiting for you is truly terrifying.

Rice Patties with Soldier Walking a Dike

We set up for the night in the corner of a dry dike tucked down in the rice paddies. Thankfully, the clouds dissipated and moon was able to give us the much needed light for our ambush. My rifle team consisted of, Jess Lopez, Dale Richardson, and Gary Gardner. Jesse was a slightly built Hispanic kid with dark, wavy hair was from Texas. He was also my RTO. He always carried out his responsibilities with quiet efficiency. We called Robert Dale Richardson "Virginian" because the good ole' boy hailed from a small town tucked away in some holler in rural Virginia. He was an accomplished story-teller, no doubt a by-product of his hillbilly

Rice Patties with Soldiers Walking a Dike

heritage. He was also a good friend. My All-Southern Boy rifle team was completed with Gary Gardner from Alabama. Tall and slender, Gary oftentimes seemed to be channeling Gomer Pyle. It was in fact this very same Gary who had tripped my flares and burned himself in the process. With his light brown hair and Vietnam tan, Gary's easy-going, pleasant personality made him a very likable guy. All three of my Southern Boy rifle team shared two common traits: courage and bravery.

Because I was so nervous that night, I took first watch which ended at 9:45pm. I had barely started to doze off when Virginian woke me up and whispered, "Hey Smooth, I think I see someone on the horizon." I rolled over and propped myself up on the dike. Sure enough there was movement. I saw 3 or 4 individuals. Despite the light provided by the moon, it was still very tough to see anything. We waited quietly until they came within 100 meters…then we opened fire. I shot the second man in line. Virginian hit the first man. The other two figures on the horizon dropped down into the darkness. We could hear a scrambling type of noise around some vegetation near the two that were shot. An eerie silence followed. Suddenly, they began to fire back and we ducked down below the dike. They were shooting from heavy vegetation. We did not know how many were out there. After the shooting stopped, we held our position for another ten minutes. We were all pretty anxious and scared at this point. Finally, I grabbed the M-60 machine gun and started blasting away at the area we suspected they were hiding. Hundreds of tracer rounds pierced the thick vegetation. After I finished spraying the whole area with machine gun fire, the silence came upon us once more. Nobody wanted to go check to see if they were still around. Since I was the squad leader, the duty fell on my shoulders. Laying down my M-60 I grabbed my M-16 and jumping over the dike, I rolled and ran to the next rice paddy dike all the while thinking to myself, *"Damn, I can get shot any second!"* I leaped over one more dike and slowly stepped toward the motionless body. It was the gook that I had shot. His comrade must have limped away or was carried by the other NVA soldiers. As I approached the body many thoughts went through my head. The enemy often booby trapped their fallen comrades by placing a grenade underneath them. If some unlucky G.I. turned the body over the grenade would explode. "Is this gook really dead?" I wondered, "Or is he playing opossum and waiting for me to get close so he can roll over and shoot me or will he explode when I turn him over?" As I continued to stare down at that motionless body lying on the ground, I pondered my dangerous situation. Was that lifeless figure on the ground daring me to turn him over? Should I or shouldn't

I? At last, I took a chance and rolled the dice. Pointing my M-16 directly at his head, I carefully flipped the body over, Yes, he was dead. To my amazement, it turned out I had killed a high-ranking NVA officer a Major! I checked the dead Major's body for information finding vital information about their company's movement through the Cu Chi Province. I also confiscated the dead major's holster and Chinese-made 9mm pistol. Both are still in my possession. But more importantly, I had "notched" a kill for Jordan. I was pleased. By this time, the rest of my men came to my side. Although we completely checked out the kill zone, the only thing we found was blood; no bodies. It was some time later we learned we had killed three of their ten man squad. Apparently one of them was captured at a later date and confessed about what happened that night.

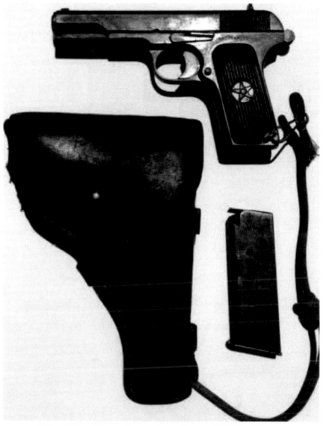

My War Trophy taken from NVA Major Chinese 9mm Pistol

As 11:00 pm approached, we were all still a little shook-up following the recent action. I reported the fire fight to the Command Post (CP). Usually after contact with hostiles resulting in dead enemy bodies, we were told to come back to the base. But after coming out to inspect the kill zone, our Special Forces captain had other ideas. He felt the night was still young and that we should go to another ambush site. After I got new grid coordinates from the Captain, we were on to our next ambush. Once again, we saddled up and made our way to our new location about one klick away. After the previous contact, Virginian had replaced his M-60 for his M-16 rifle. We moved along the dike cautiously until we reached our destination. The men were still pumped up from the earlier action and kills. They were chattering among themselves about my impressive M-60 skills. "Hey, didn't Smooth look like G.I. Joe with an M-60 tucked under his arm?" laughed Virginian. "Yeah, he stood up and started blasting away," Jesse chimed in. As Gary chuckled, he added, "Then he switched to his M-16. Smooth was tumbling left then right with his M-16 like John Wayne." They were having just a little too much fun at my expense; little did they know I was shaking in my boots. I finally had to put an end to their good-natured fun, "Okay guys, you all need to settle down," as we continued on to our next ambush site.

Everybody was still annoyed we had been sent on another ambush. After some minor complaining by

Me holding an M-60

everyone, their high-energy level subsided and we settled down for the business at hand. We set up again in the corner of a dike with limited visibility. Since I was still in no mood to sleep, I took the first watch. My mind was replaying my earlier kill. I was still mulling over what "might" have happened to me if the other soldiers would have been hiding in the bushes waiting for me or if the Major had still been alive. I counted my blessings and tried to refocus on guard duty. About a half hour later, I looked down the dike. From my seated position in the corner of the dike, I spotted a silhouette. The rice paddies had been harvested, so there was no longer any rice growing to obstruct my vision. Without a doubt it was the enemy! There were 10 to 12 men in full military gear; obviously NVA soldiers. From their angle of approach, they could not see me or the others. I quietly rotated my body to face them. I lifted my weapon and focused on their point man. The point man saw me as I positioned my weapon towards him. His rifle was pointed directly at my head from about a distance of five feet! Before he could get a round off, I opened fire. Then, the shit really hit the fan!

The fire fight was on! I kept firing down the dike knocking numerous NVA off of it like mechanical ducks at a carnival shooting gallery. Although my sleeping men were briefly stunned by what was happening, they quickly arose and started firing back. As the bullets whizzed past our heads left and right, it occurred to me we were outnumbered. And to make matters worse, we were trapped in our corner with no way out. I thought to myself that the enemy didn't know there are only four of us. It was dark. Since we were all against the dike wall, the lack of background obscured their visibility. I was positive we killed or wounded most of their men. I knew I had personally hit anywhere from four to six of them. After about 10 minutes of exchanging fire, I told my men to grab their hand grenades and throw them over the dike in the direction where we thought the enemy was entrenched. I told Gary, "Toss your grenade then run to the other side of the dike. It's only about twenty meters." He did exactly what I had instructed him to do. Now it was Virginian's turn, he took off and sprinted to safety on the other side of the dike with Gary as we opened fire and threw a couple of grenades. "Lopez," I yelled, "You're up". "Smooth, I'm not going until you do," he replied. "I appreciate it buddy," I said. "But you got the radio and we can't afford to lose you."

Jesse Lopez
Radio Transmitter Operator

When Lopez got up to make his break, I opened fire and threw the last of my grenades. He reached the others without injury. Now, I was all alone. There I was with no backup. I was on my knees leaning forward on the dike. I was shooting at anything that moved. Emptying one clip and quickly reloading, I kept firing in all direction into the darkness occasionally seeing a shadow of the enemy. I had thrown all my grenades and now it was time to exit the firing zone. "Don't shoot," I yelled to my men. "I'm coming over." There was very little light and I didn't want anybody to get anxious about any movement heading in their direction. I ran and jumped over the dike successfully. Grabbing the radio, I called the command post about our situation. "We're outnumbered and need flares", I yelled. I gave the grid coordinates and told them we were taking small arms fire from 10 to 12 NVA. I moved closer to the enemy crawling up on a dike. When the flare lit them up, I wanted to be in position for a look or possibly a kill. Unfortunately, the artillery put light in the wrong grid. It was right on top of us! Thank God it wasn't artillery rounds. We would have been killed by friendly fire! But now I was a sitting duck.

Radio Transmitter Operator

Hugging that dike like it was my lover; I didn't move a muscle. As I looked down at my sweaty arms, to my horror, I was being attacked by the biggest ants I had ever seen. They were both red and black ants! They started biting me like they were starving. My only option was to stare at them nibbling away while at the same time keeping an eye on the direction of the enemy. I couldn't wait for the flares to cease so I could move back behind the dike. When the last flare went out, I quickly jumped back over the dike and frantically brushed off the ants.

I called the artillery back and told them to discontinue the flares. By this time, the Captain was on the horn telling me to hold my position, "Help is on the way," the captain insisted. We continued to take fire and return it. However, it seemed less fire was coming in than before. It appeared they were retreating. Suddenly, we looked behind us and there was Captain Fox in an armored vehicle with his lights on. He was jumping the dikes and was followed by thirty men. Those approaching headlights piercing the darkness were a beautiful sight! We stayed where we were and he ordered the platoon to go in the direction of the enemy. Safe at last and nobody was hurt. The only injuries I sustained were bites on my arm inflicted by ferocious, flesh eating ants. We returned to our perimeter about 1:30 a.m. Thankfully, the Captain allowed us to go back to the base and not out on another ambush. Back within the safety of the perimeter, we tried to settle each other's nerves. We had triumphed and survived two perilous contacts with the enemy in one night. December 20, 1968 had indeed been *"a night to remember."*

Virginian & Me

Lopez, Virginian, Gardner, & Me

Virginian, Gardner & Me

A Walk Through the Valley of Death

CHAPTER SEVENTEEN

My rifle team was called in the next morning by a lieutenant to explain what had transpired during both of the ambushes. Later that day, we were also asked to go to a nearby village. It appeared someone had shot off the tops of their huts. I had accidentally taken out a few thatched roofs while blasting away at the enemy with my M-60. Whoops! Well, shit happens. When we saw the hooch roofs dismantled, my men looked at me and we all started laughing. Virginian sarcastically remarked, "Nice shooting Smooth. We'll have to keep that M-60 out of your hands from now on." At the village we also saw 8 dead bodies laid out. One of the dead men was about 6 feet tall. He was definitely Chinese. The lieutenant who had heard our report of the action asked me to walk with him to see one specific body. "This particular soldier stuck out because of the report you gave me this morning on the ambush," he said. "He was their point man and his trigger finger had been shot off!" I was speechless. This was unbelievable! "Did I really beat him to the draw?" Apparently, I had managed to dodge death for the fifth time. Out of the corner of my eye, I thought I saw Lady Luck winking at me. "Jordan may you rest in peace,"

I said to myself, "I have vindicated your death with eight more kills." It may seem like a sentimental cliché keeping my promise to Jordan. In order to truly understand those sentiments, a person would have to hump several klicks through the jungle in my combats boots.

My body count was somewhere around 13 at this time. My teams' numbers were about 20 kills. Why is that important? There were rumors that the squad with the highest body count (kills) would be allowed to go to the Bob Hope Christmas Show. The selection process was a bit morbid, but hey it's Bob Hope! We went on two more ambushes without incident. Then, we got the good news; we were selected to go to the Bob Hope Show!

We were taken by chopper back to an Air Force base where Bob Hope and Ann-Margret were performing. We sat in the front row. As we took our seats all eyes were on us. We were being sized-up by those in attendance. Everyone was looking at us strangely. We had just come from the bush, so we were heavily armed and didn't smell very good. The audience at this show was mostly Air Force and military, clerical personnel. They had not seen or been near any combat action. The Air Police (AP) asked us to unload our weapons. "Sure thing," I said as I put my rifle on safety. "There, it's all taken care of." The

Bob Hope & Ann Margaret Dec. 1968

AP's just shook their heads and walked away. It was a great show and just what we needed to bring us back to the real world for a while. Bob Hope was hilarious and the gorgeous Ann-Margret wowed us in her '60s era mini-skirt.

We were allowed to hang out a couple of days before being sent back. During our extended stay, we enjoyed real hot food and got to watch a movie. The facilities were amazing to us, housing, real beds, air conditioners, a daily change of clothes, real restrooms and yes, even real showers! Here, we had cold beer. Normally the beers out in the field were hot. All these things people take for granted was a luxury to us. As we headed back to the fire base, I thought about my family. Maybe they would see me on TV at the Bob Hope Show since it was taped prior to Christmas. I still have very fond memories of that show (mostly of Ann Margaret). I am still in awe of Bob Hope and Ann Margaret for their dedication to their country and to the men and women at war. I applaud these entertainers for coming into harm's way to put a smile on our faces.

A Walk Through the Valley of Death

CHAPTER EIGHTEEN

During this time, I found out the United States had begun the draft lottery again. The first one was at the beginning of World War II in 1942. In December of 1969, there were 850,000 men eligible for the draft between the ages of 18 to 26. The process for the lottery was much like pulling numbers for Bingo. The bowl contained 366 birth dates. Each date, randomly selected, received a corresponding number beginning with the number 1. Numbers 1 through 122 were going to be drafted. Numbers 123 through 244 were "on hold". These numbers were on standby in case more soldiers were needed. The remaining numbers 245 through 366 would not be drafted. Those lucky bastards were free to go on with their lives. September 14th was selected as the number one. Therefore, all men born on that date, between 18 and 26 years old, were going to be drafted. There were 283,333 men selected in that first group. Men born on my birth date of April 12, 1947 had the number "346". Had I not opt for the volunteer draft, I would have been exempt!! Go figure!

We returned to the field and continued to go in and out of villages. After a

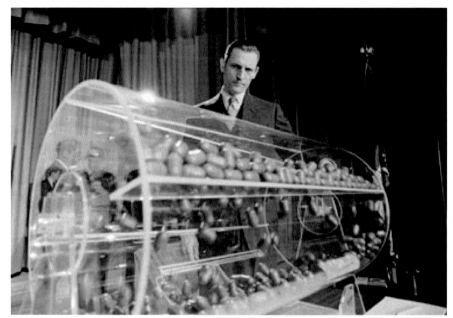
Capsules containing birthdates of eligible draft age men

Search and Destroy, we settled in a quad about 50' by 75' to eat and sleep. There were about 15 to 20 Army of the Republic of Vietnam (ARVN) troops there mingling among us as well. The ARVN were renowned for their lack of combat prowess. They were infamous for running from battles rather than engaging the enemy. We joined them around a fire and drank some liquor called *sploe*. Sploe was light, amber-colored liquor like tequila, but packed a stronger punch. It was getting dark, so one of the ARVN troops grabbed the bottle we were drinking from and poured it into the lanterns. The lanterns lit up like flood lamps which gives you an idea of sploe's incendiary potency. We continued to party late into the evening. It was time to get some sleep. We rolled over with our poncho liners, weapons and passed out. The following morning the ARVN troops were gone. A subsequent intelligence report informed us we had "slept with the enemy." We had been drinking with VC "wolves" dressed in ARVN "sheep's" clothing! We asked ourselves, "Why didn't they kill us in our sleep?" "Were they too hung over?" We never found out. Besides, it's better to laugh about it now than to think what might have occurred during this truly surreal encounter.

A village near the city of Saigon.

It was not uncommon for enemies to stop fighting and join together and break bread. During the American Civil War, combatants often stopped hostilities for lunch. British and German troops shared holiday cheer during Christmas truces during World War I. Hell, they even played a soccer match against each other! But after that brief pause in hostilities, they went back to their respective trenches and resumed the slaughter that was The Great War. Future historians of the Vietnam War might not put this in the same category as those other stories. But I think our "sleeping with the enemy" anecdote fits nicely in that category. Fortunately for me and my men, life went on.

Another village near Saigon.

We headed into some farmland looking for a reported cache of weapons and VC. As we humped over the land, I noticed some tall plants that looked vaguely familiar. Upon further inspection I discovered to my surprise that it was wild marijuana plants. I decided to give the trailing platoon, about two klicks away, a special treat. With my handy dandy, zippo lighter I set the weed on fire. The wind was blowing conveniently downwind right towards the platoon. We never found any weapons or VC, but the last platoon thanked me and said it was "really groovy man!"

Later on, we were moved to a home base close to Saigon. Carl Linze was a very muscular 6 foot plus, two hundred forty pound fellow from Montana. All that bulk and muscle made him an ideal man to handle an M-60 machine gun. One evening around 6:00 pm, Carl had joined Dale, Luna and me into downtown Saigon. We were listening to some American music in a nightclub; drinking beer and just relaxing. A few tables away from us, we became aware of Carl trying to resolve an "issue" with a local Vietnamese patron. It wasn't long before we overhead them yelling at each other. It seemed like a comical confrontation as Carl towered over this much shorter guy. Suddenly, Carl picks up this pint-sized 90 pound gook and throws him through a window. Five other guys started to charge Carl. Naturally, we went over to help out our buddy. A barroom brawl broke out. It felt like being at home with my two brothers. We were throwing these little guys all over the place. Then all of a sudden one of them pulled out a gun and shot Carl in the chest. The shooter ran outside and jumped on a scooter and drove off. I chased him down the street until he started shooting at me. Since I was unarmed, which was rare for me, it seemed like a good idea to stop the pursuit. An ambulance came and picked up Carl and took him to the

Carl Linze

local American hospital. He was conscious, but blurry eyed and not very coherent. It was difficult finding a ride to the hospital. It was midnight before we finally arrived. The doctor came out and gave us Carl's prognosis, "Your friend is in critical condition," he said in a cold, clinical, somber tone of voice, "But he's going to make it." He showed us what had saved Carl's life. His metallic lighter had slowed the penetration of the bullet that would have otherwise hit him directly in the heart killing him instantly. Carl owed a debt of gratitude to the good folks at Zippo.

Party and Brawl in Saigon Bar

CHAPTER NINETEEN

Shortly after that, we found ourselves back in the field. We were on our way to another hot area on a Search and Destroy. Despite not making any contact with the enemy, everything seemed a little tense. I couldn't shake the feeling that someone was following us. We stopped for a C-rations lunch break. Spaghetti, pound cake and peaches were my favorite selection off the C-rations menu. I used my handy, dandy, army issued P-38 (can opener) to enjoy my meals. Apparently, the gooks loved spaghetti too. So, I left a specially prepared gourmet can of spaghetti. After pulling the pin on one of my grenades, I delicately placed it in the empty can and then carefully closed the lid. Bon Appetite. If some unfortunate spaghetti loving enemy troop opened the lid, the handle would pop and BANG! You have spaghetti ala gook. We walked about a half a klick away and KABOOM! In the deadly game of booby traps, it is all fair play. My instincts had proved correct; someone had been following us.

That night the captain let us go back to a safe fire base. After setting up in bunkers, we all relaxed. The welcome downtime was taken up writing letters and smoking dope. I called my pipe "The Destroyer" because when you smoked marijuana with it, it would destroy you! The only time my squad was allowed to smoke was when we were not going out for a few days, otherwise, no other drugs were allowed.

We roamed the base feeling pretty relaxed and we all had the "munchies." We located some cheese and crackers and sat down to eat our snack. The supply sergeant set up a lean-to screen and a projector outside the bunkers so the company could watch a movie. After the movie we returned to our bunkers. Some of the men continued to party and lit up another doobie; others went back and wrote letters home. I could only imagine what was being written in those marijuana influenced letters. Often times in my own doobie influenced letters they included long, rambling passages about "peace" and other psychedelic notions of the Beatles and "Strawberry Fields." I specifically remember getting wrapped up in the notion that while people in the United States were safely at home in their cozy houses with children playing in "strawberry fields," children here were being

Me and the Destroyer

injured and killed. People in the States had no idea what real horror was occurring. Like the song says, "Living is easy with eyes closed".

I'm holding a picture of children frolicking in a field over punji sticks as an ode to the song, "Strawberry Fields Forever".

We were suddenly interrupted by the entrance of our lieutenant into the bunker. BUSTED! The whole place was thick with the pungent aroma of dope. He looked at us suspiciously, but we reassured him, "Everything's is just fine. Don't worry about it!" He just nodded in resignation and left. We all looked at each other and busted out laughing. The lieutenant was no dummy; he knew

Close up picture of children frolicking.

exactly what was going on. Days later, I would receive a letter back from my sister expressing some concern about those letters I wrote. She would comment, "You're all over the place! I don't know what you were trying to tell me; are you okay?"

The next few days were pretty mellow. One night a lieutenant came to see me and asked me to report to the Colonel. I immediately went to his well secured bunker. "Have a seat soldier," the colonel said, "I have a couple of questions for you." I sat down in a chair next to his desk. He said, "I have reviewed your records and it has come to my attention that you have a college background and that you have shown bravery in the field." I was shocked to hear anyone knew who I was or cared what I had done in combat. Then he asked me, "Fernandez, how would you like to become a Warrant Officer and learn how to fly choppers?" I was excited at the offer. I had always dreamt of flying a chopper. BUT, there is always a catch. "You'd have to go back to the States for six months of training," he continued, "After that, you'd be shipped back to Nam for two tours of flying a gunship." It was a tempting offer. However, I just couldn't see myself in the Army an additional four years. The mandatory two tours of duty flying a gunship in and out of "Hot Zones" in Nam also had its deadly drawbacks. "I appreciate you thinking of me Colonel," I said, "But I'm going to have to respectfully decline your offer."

CHAPTER TWENTY

January/February was typically Monsoon season in Vietnam. During the heavy downpours, we would take showers outside in the rain. Like clockwork, the rain would start at 3:00 pm. and it would only last about fifteen minutes. In anticipation of the rain, you would see 50 or more naked guys with soap in their hands waiting for the inevitable downpour. Thanks to the heat, we would dry off in about ten minutes. Also during these days of being in the rear we would have some fun playing football. I

Vietnam Footballers

remember calling a play telling my wide receiver to do a corner/post pattern around a pungi pit. The pit was about 6' in circumference and 4' deep. My receiver was wide open because the cornerback had no chance of jumping the pit. Touchdown! For all of us, it was a much needed relief from the reality of combat.

It was the middle of the day and the lieutenant came to me and said, "Smooth, time for you to take a five day R & R, (Rest and Recuperate), you are going to Hong Kong in the morning." That was music to my ears. Without saying a word, I grabbed my gear and went back to the air base for the night. In the morning, I boarded a commercial flight with a buddy, Ben Shimpff, born and raised in San Francisco and currently living Southern California. We landed in Hong Kong. We immediately split up and planned to meet at a night club later. I stood there all alone in a city that looked like downtown San Francisco. I checked into a hotel and exchanged my US dollar for yen. The rate of exchange was 360 yen for every US dollar. With all the yen stuffed in my pockets, I knew a good time lie in my future. Hong Kong was famous for sight-seeing adventures, a ferry boat trip to Kowloon, and a distant glimpse of the Great Wall of China. Hong Kong was also famous for making custom suits. Feeling like a millionaire, I had two custom made suits to take home with me. We also heard about the colorful night life. Ben and I met at a nightclub. I have no recollection of that night, but I was later told we had a wild time!

One of the highlights of this trip was that I met a woman named Sherri. She was a beautiful, tall, slender nurse with shoulder length blond hair and blue eyes. She was a nurse who worked at one of the MASH units in Nam. She was from Santa Barbara, kind of a beach gal who liked to surf. She had her college degree and said she would continue nursing upon her return to the states. We ate dinner and went dancing. After sharing some of my combat experiences and my persistent nightmares, she was overwhelmed, but sympathetic. Some of my behavior patterns I had acquired in Nam were odd. I chose to sleep on the floor because the bed was too soft. Also the height of the bed exposed me to the window making me feel vulnerable. In my mind, there was someone who wanted to shoot me from outside. Sherri understood and didn't judge or question any of my bizarre behavior. Before I went back to "Hell," she gave me a sterling silver charm for good luck. It was quite thoughtful and I was grateful. The next day I met Ben at the airport and we boarded the plane back to "Flower Land." One man from our company did not return to the plane and went AWOL in Hong Kong. To this day, I never heard if they found him.

A Walk Through the Valley of Death

CHAPTER TWENTY-ONE

Our squad went out on the road again. This time we were looking for land mines, specifically designed to destroy our tanks. This involved four soldiers operating landmine detectors. They rotated across a twenty-five foot road and have several tanks and personnel carriers behind them. In war movies, it is always depicted that a soldier detecting mines can be shot at any time because they are out in front, totally defenseless and exposed. But, what is not shown in those movies is that there are point men out in front protecting those mine sweepers. Since it was unlikely that our body weight would set off the mines we were put out in front. That was the mission. Walking down the road, knowing you could be shot by a sniper at any time was a pretty unsettling thought. The possibility of an ambush was reduced by the presence of the tanks, even though they were about 100 meters behind us. The real danger was snipers. Fortunately, we once again beat the odds and the tanks got through. No snipers. Several days on Search and Destroy missions resulted in no contact. We were all fine with that. As draftees, our real "mission" was to survive and get back home in one piece. When I left

Vietnam in 1969, I felt we were winning the war. The bombing in Hanoi was bringing the enemy to their knees and to the Peace Table. Although at the current time there were rumblings among officers and enlisted men that we were just "spinning our wheels." Sentiments were growing that the U.S. government was handcuffing our war effort by not utilizing our vastly superior military power.

The Vietnamese have been fighting against foreign invaders for thousands of years. Over generations of guerilla warfare, they had developed tactics and built an intricate network of structures to aid in their efforts. One example is the Ho Chi Minh Trail. It was a vital supply route made up of trails running not only through Vietnam, but also Laos and Cambodia. Since Laos and Cambodia were neutral countries, we were not "legally" allowed to conduct military operations there countries. Our Secret War in Laos, Nixon's Incursion of Cambodia and bombing the Ho Chi Minh Trail proved ineffective. Some military big shots even talked about using our nuclear weapons to turn the tide of the war. But it became apparent when the Paris Peace Talks began that the politicians had taken command of the war. During their negotiations with North Vietnam, we were not allowed to shoot at the enemy! The NVA and VC never got that particular memo. Imagine fighting a war with one hand tied behind your back while not being able to initiate fire when you see confirmed enemy. We were only allowed to return fire if we were fired upon. We are out on another patrol. The search and destroy mission was taking us down a trail towards a rice paddy. We noticed the rice paddies were heavily flooded, so we walked on the dikes. From my position on point, I spotted something moving about five feet in front of me. I stopped, and then took another step. Something big and fat in the grass was slithering in front of me. I couldn't believe my eyes. It was the biggest fucking snake I'd ever seen. In that brief, horrifying moment, I mentally flashed back to a high school track and field meet. I see myself running down the track toward the long jump pit. My heart is pounding and I can hear my breathing getting faster, deeper and stronger. Now I'm back by the dike with the monstrous, nasty snake looming in front of me. My old track and field skills instinctively kick in. I was back at the track meet. I hit the take off point perfectly. I was airborne, defying gravity and every other law of physics. I'd made the longest jump of the day! The cheers from the stands were deafening. Flashing forward in time again, I discovered my incredible one-legged leap into the air and over to the other side of the dike clearing that big black venomous snake! I had thoroughly amazed my squad. "Jesus Smooth, you looked like Bob Beamon, (U. S. Olympic long jump gold medalist) flying in the air like that." "Sucker looked like he had wings, didn't he?" I had made a clean jump at an angle of about 8 feet to the other dike.

"Snakes, why did it have to be snakes?"---Indiana Jones, Raiders of the Lost Ark

I was not the only one who had issues with unwanted creatures in Nam. Oscar, one of my buddies from the armored division and former roommate, told me a story of a scary "critter" encounter he had experienced. While on night ambush with his squad, he could hear movement throughout the night, Oscar began. "I'm telling you man, the whole night was really freaky," he said. "I was sitting there with my rifle on guard duty when I smelled something really bad…like shit." It was common knowledge among those of us who had spent time crawling around in tunnels that the telltale odor of human feces meant there were VC in the area. "Then from out of the trees, something jumped out", he continued, we all starting firing our rifles at an invisible enemy. Two attackers jumped on my back. I spun around and they flew off," recalled Oscar, "It was a couple of MONKEYS!" That's what he heard and that's what smelled so bad. He said it scared the shit out of him. Ever since that experience, he is afraid of monkeys.

CHAPTER TWENTY-TWO

The time was approaching for me to exit from "in country." I was getting "short." Getting short means you have less than 30 days left in Vietnam before going home to the "real world." As my time in Vietnam was winding down, I found myself on the horns of a dilemma. On the one hand, I was having second thoughts and reservations about the dangerous assignments for which I had volunteered. But on the other hand, I felt very confident in my ability to take the risk if that would help me and my men survive. So I found myself asking, "Do I trust someone else less qualified to walk point? Or do I take the risk and do it myself?" I decided to be more cautious and selective in the actions I would take from here on out. As we go out on another night ambush, I'm concerned. The area is "hot." Suspected Viet Cong had been spotted in our grid. There is no moonlight. We're walking on dikes that seem higher than usual. My six foot frame is now elevated higher making me an easy target…so, yes I'm very concerned about our safety. Luckily, nothing happened on the way to our ambush site. After we set up for the night, everything seemed okay. However, I stayed alert, keeping all my senses sharp for anything out of the ordinary. Because I was again too "wired" to sleep, I took first watch. After my shift, Luna took over. "Don't fall asleep," I said firmly. "Make sure you stay alert." About two hours into his watch, I looked over at him and noticed he was sound asleep! I grabbed him from behind and placed the "business end" of my hunting knife to his throat drawing a speck of blood and warned him, "Next time the blade goes deeper if I every catch you asleep again!" Luna had compromised our safety by falling asleep. Perhaps my "teaching" method was harsh, but under the circumstances I felt it was justified. Luna got the message. We all survived another night ambush mission without incident. We continued search and destroy missions for the next few weeks without anything big happening. Our next mission was to board a boat with the "River Rats."

River Rat Patrol Boat

These guys were crazier than me. Armed with .50 caliber and M-60's machine guns, all the crew members carried M-16's as well. The "River Rats" would patrol the waterways close to villages looking for anything suspicious. They were also used to transport soldiers. Additionally, they carried out night ambushes, hit and run raids, and special intelligence.

The dense vegetation growing near the riverbanks provided a perfect hiding place for the enemy to shoot rocket launchers at these river boat patrols.

NVA with Rocket Launcher

I couldn't wait to get my boots back on dry land! We pulled over to one of the suspected VC villages. From there, we went on foot patrol. We captured a gook running down a trail. He was interrogated by an interpreter. The captured suspect claimed he knew nothing. One of our officers thought the prisoner had information he should share with us. After asking the prisoner one more time about his unit and their location, the interrogation turned to a very serious confrontation. The prisoner was put into the shallow part of the river, the interrogator continued the questioning. There was no response. Then he had the RTO hook up the transmitter wires to the prisoner. After a few jolts of electrical current, the gook's memory came back. He gave directions where his Viet Cong platoon was headed. The Captain called in "Puff the Magic Dragon." These Air Force AC-47 gunships were armed with 7.62mm mini-guns. With their mini-guns firing down a hailstorm of lead, they could kill everything within the area of a football field in less than ten minutes.

Captured VC & Taken for Questioning

The AC-47 gunship earned its nicknames of either "Puff the Magic Dragon" or "Spooky" because of the smoke and flames that would spew out of the mini-guns mounted on the fuselage giving it the appearance of a "death dealing, fire breathing dragon." Puff took care of the problem. The captive also told us where a cache of weapons was hidden along with rice sacks and medical supplies. He was then taken in as a prisoner and no further harm came to him.

Puff the Magic Dragon AC-47 Gunship

CHAPTER TWENTY THREE

I was now down to my final 5 days! We were back at a relatively safe fire base and all seemed good. My understanding was that I would finish my tour of duty back at the base. By this time, I had received a number of medals; the highest was a Bronze Star, and a second Bronze Star with a cluster and "V" device. Both medals had been awarded to me for gallantry. I didn't care about all these medals; I just wanted to go home. The last thing I wanted was a Purple Heart or to go home in a body bag.

According to Army policy, a psychological evaluation was mandatory 30 days prior to returning to the States. The stated purpose of this decompression time was for "readjustment" out of the jungle and back into the "world" aka The United States. It never

Me, Virginian, Lopez receiving bronze star

Bronze Star w/V-device

Bronze Star w/V device & Oak Leaf Cluster

happened for me. The Army's failure to fully implement this "readjustment" policy resulted in thousands of Vietnam Vets coming back home to the States after serving their country suffering from PTSD.

With only four days left to go, my Staff Sergeant approached me and said, "Smooth, we're going on an Eagle Flight." That was the last thing I wanted to hear. An "Eagle Flight" is a large air assault of helicopters in a hot zone. So I got my squad together and told them we were the lead chopper.

At the time, I didn't know we were going into Cambodia! Cambodia was a neutral country; we had no legal authority to enter in that country. Nixon had made an agreement with the leader in Cambodia in 1970. But it was common knowledge that many Special Forces units and long range patrols had been in both Laos and Cambodia. We mounted up and got into our

Preparing to go into Cambodia

choppers. Our landing zone was supposed to be "hot." However, upon landing, we did not encounter any hostiles. We stayed for a few hours; got back on the choppers and flew back to base. Getting back without incident was a huge relief to me.

The next day the Staff Sergeant said we were going out again. My squad, once again, would be the lead chopper. He also had some other news for us. There was a fresh "cherry lieutenant" in our platoon who would be going along too. I have three days to go and a rookie lieutenant is in charge! Needless to say, I was getting very anxious. I told my squad, "Get on the chopper and stay alert!" We were heading back into Cambodia. The landing zone was lit up with "red smoke." That signaled that the area was "hot" with hostiles in the vicinity. Our pilot got scared refusing to land the Huey. This pilot was the exception to the rule. Most chopper pilots had balls of steel. This pilot told us, "Jump out of the chopper!" He had to be kidding! We were hauling 60 to 80 lbs. of gear on our backs. A sure way to break a leg is to jump out of a hovering chopper onto a wet ride paddy! We were about 7 feet from the ground. When we hit the mud paddies we sank up to our thighs. We then encountered small arms fire. I looked at the rookie lieutenant and could see he was about to shit in his pants. "Smooth," he said shaking with fright, "You take over the platoon." Yes Sir! I did so without hesitation. Nobody had any confidence in him. I ordered the squad leaders, "to pull themselves out of the mud and in stagger teams get to dry land". We scattered behind high vegetation for cover. A small fire fight ensued, but we couldn't really see the enemy due to the heavy elephant grass. Elephant grass is tall with razor sharp leaves. If it caught you at the right angle, it would cut your arms or face. I usually pulled my sleeves all the way down and buttoned my cuffs to prevent being slashed. The NVA were outnumbered and quickly retreated. The lieutenant asked, "Smooth, what should we do?" The lieutenant was asking advice from an extremely "short" guy. Less than 2 days left. So I lied, "We should hold our position. We don't want to give our location away", I said. "In case they have mortars." My normal impulse would have been to pursue the bastards. The squad knew what I was doing and laughed to themselves quietly. The rookie lieutenant agreed and we stayed there for a couple of hours. Then we called the choppers to pick us up and return us to base.

With only 2 days left, Sarge told me we were going into Cambodia again! Unbelievable! "Jesus Sarge," I protested, "I'm going home tomorrow!" "That's not my problem," he answered. He'd left me no choice. "I'm not going," I said. Then he said, "What did you say?" I repeated, "I'm not going!" He reported me to the captain. A few minutes later, the captain had a little chat with me. He said, "I was told you are not going on the chopper assault?" "That's correct sir, because I'm going home tomorrow," I explained. "If you don't follow a direct order, I'll have to court martial you," he said. "You have to do what you think is right, and so do I," I replied. He informed me that they were planning on increasing my rank to sergeant if I went on this last mission. I refused to negotiate my life for a higher rank. To me this was similar to being offered the Silver Star to go down the tunnel. It was also the same feeling I had about the situation with Billy when he was killed in the tunnel. This was not a good idea.

I left our camp and caught a truck back to the rear base. After picking up all my gear, I headed for the airport for processing out. I found out later that there were a number of casualties on that last chopper assault into Cambodia. One individual, who died in action, received the Congressional Medal of Honor, Felix Conde Falcon, the highest award to be given to a service man/woman. Felix took out three heavily armed bunkers protecting his platoon and killed several of the NVA before getting hit and killed by a sniper. They hit a battalion command post with a 1,000 NVA. Fighting was fierce and there were many casualties. A number of tactical Artillery and Air Strikes bombarded the enemy sites. I had to ask myself, why did these brave warriors have to die or get injured in a country in which we did not belong? Knowing now what happened in Cambodia, I cannot help questioning my actions. What if I would have gone on this last chopper assault? Would my last day in country really be my last day on this earth? Would I have made the ultimate sacrifice or would I be spared one final time?

As I approached the airport, I was expecting the Military Police to be there waiting to arrest me. Nevertheless, I'd still be allowed to go home. There would more than likely be a court martial, dishonorable discharge and some time in a federal prison. But as those thoughts were going through my mind, I kept thinking about Billy, Jordan and the guys in my squad. My instincts were correct. I had done the right thing for my men and myself, regardless of the consequences. My original goal when I arrived here was to stay alive and get back home. I had done it. I was going home ALIVE. It was time to face the music, my final processing was about to occur. As I approached the administrative desk, the sergeant never looked up at me. "You didn't get a court martial," he informed me, "They're giving you an Article 15." An Article 15 included a reduction in rank and would go in my personal file. I thought, "Great!" Then abruptly without warning, the processing clerk tore up the Article 15 papers throwing it all in the trash! What the Hell was going on? I was shocked. "What are you doing?" I asked. 'Smooth," he answered, "You saved my life and this is the least I can do to repay you." I was stunned! The sergeant looked up at me. It was Mickey McKay, the rich kid from Cape Cod! Yeah, I guess I had saved his life. For me, it was all in a day's work. It was one brother in arms looking out for another brother. The Old Time Vets had taught me that. It was just something you did without thinking about it. Apparently, Mickey had been sent back to clerical. No doubt his family was able to pull some strings and get him out of the action. Mickey said, if I was ever in Cape Cod to please call him, "My home is your home Smooth," he said and that made me feel good. "How ironic," I thought to myself, as I left the processing center. After all I've been through in combat, escaping death, that my final exit and saving grace from the "Flower Land" came down to a clerk tearing up a piece of paper and tossing it in the waste basket. Swish! Nothing but net!

I boarded the "Freedom Bird" bound for home. It was the happiest day of my life!

EPILOGUE

On the day that I returned home from Vietnam, I took off my uniform, threw it in the back of the closet and suppressed my memories of all that I had endured.

The G.I. Bill proved invaluable to me, I immediately enrolled in school and I successfully completed my college education at San Jose State University. My goals of becoming a teacher, athletic director and basketball coach at the high school level were at long last realized. Ever since, I have been teaching and coaching basketball. It's always a thrill and very gratifying working with young student athletes. Teaching and sports has always been a major interest and part of my life.

As a life-long fan of the Dallas Cowboys, Oakland Raiders and San Francisco Giants, I've followed them faithfully through their glory years and not-so-glorious years. Golf is another passion of mine. Whenever I can fit it into my busy schedule, I look forward to getting out on the links to play a round of golf.

February of 2008 proved to be a pivotal and decisive date for me. Prior to that time, my involvement in the movement to create a Vietnam War memorial to honor the 142 young men from San Jose killed in Vietnam was still only conversational and theoretical. I knew the memorial was a great idea and long overdue. So in early 2008, I made the commitment to dedicate myself to making the dream of a Vietnam War memorial a reality. Thus, the San Jose "Vietnam War" Memorial Foundation Inc. www.sjwarmemorial.com was created.

Early on in its infancy, the Vietnam War Memorial project was in a shape-shifting phase. It was changing from just a nebulous, well intentioned idea into a project with many facets and dimensions. There were endless meetings with members of the San Jose City Council authorizations, allocation of space and other "red tape" matters. The five board members of the Foundation each brought an expertise that proved essential to completing this project.

During this long journey, we started getting positive involvement from various firms and individuals offering support. Erecting a monument in downtown San Jose is an enormous logistical undertaking requiring a wide range of knowledge and skills.

Fundraising for the memorial project consumed a great deal of our time. We found ourselves making endless presentations before groups of prospective donors. There were venues ranging from private offices to a Mercedes-Benz showroom. We were making the case for the need for a Vietnam War monument honoring those sons of San Jose who had died in action. It's said public speaking is the scariest experience an individual can endure. Speaking in front of a group of strangers, according to so-called experts, is more nerve-wracking and harrowing than being in combat. Although I never had difficulty speaking or lecturing my students in my Social Science classes throughout my 32 years of teaching, I am not 100% comfortable speaking in front of a crowd. But I'd rather give a speech to the City Council or Rotary club than going down a tunnel. I don't think those experts ever had the pleasure of nearly falling into a punji pit or crawling down into a tunnel with only a pistol and flashlight in total darkness searching for a hidden enemy.

In 2012, our monument was complete and unveiled on March 30th 2013. Our monument is located in downtown San Jose in the Guadalupe River Park at Confluence Point. This day has been designated as "Welcome Home Vietnam Veteran's Day". Every year we return to the monument site on March 30th to read all 142 names. The Sons of San Jose will never be forgotten. The memorial is my way of also honoring and remembering all those who served, especially my fellow brothers in arms. 82nd - All the Way!

Photo by Sandra MG. Fernandez

Photo by Sandra MG. Fernandez

Photo by John Lozano

Photo by John Lozano

A Walk Through the Valley of Death

ACKNOWLEDGEMENTS

Perseverance helped me achieve my goals in life. But, I did not do it alone. I would like to thank my family for their ongoing patience and support. When I returned from Vietnam, there was no re-integration process. A soldier just came home and was expected to "move on." Dealing with me during that time, I am sure was difficult. Luckily, my family who loved me just let me be. I chose to pretend I could forget about the war and rarely spoke about my experience. Unfortunately, hiding from these demons only caused other issues. I regret the mental anguish I must have caused for two ex-wives. I doubt that their forgiveness will ever come and that will always be something I have to live with.

This book is about coming back to the "world" from a very dark place in my life, yes, much like the tunnels I entered. After 30 plus years of avoiding the subject, this book was written to rid myself of some of my nightmares and uneasiness in my walk of life. It does not resolve all of my PTSD issues, but it has helped to heal them.

My wife, Sandra, is the person who was able to put a handle on my idiosyncrasies, not by justifying them, but by listening to me. She encouraged me to talk about my moments of depression, anxiety and flashbacks. Sandra has been my strength to get through this book and life, as there were many a time I wanted to quit because of the pain of remembrance of the days in the "Jungle." I am blessed for her support and wisdom to complete this book. Her input and patience to help write, edit and enrich the content of this book was invaluable. I chalk it up to love and compatibility.

Roberto "Bob" Leal was another strong contributor to this book. He too was a positive force and encouraged me to finish. He put in endless hours editing the book, added relevant knowledge and anecdotes and for that I am grateful. Lastly, Craig Luther, an author in his own right, also assisted with the editing and for that I am thankful.

I would of course need to acknowledge all of my children for they too have had to endure my many mishaps in my life. My children from oldest to youngest, Dennis J. Fernandez Jr., Suzanne E. Fernandez-Hussla, Shawn TG. Fernandez, Kellie M. Fernandez, and Zachary D. Fernandez, will always have my unconditional love.

GLOSSARY

AIT: Advanced Infantry Training 2nd 6 weeks of training after initial basic training

Agent Orange: An herbicide and defoliant chemical used to kill plants and trees. The chemical has traces of dioxin which has caused health issues to those exposed.

AK-47: Russian made Assault rifle used by the NVA and VC

APC: Armored Personnel Carrier

ARVN: Army of The Republic of South Vietnam allies with the United States

AWOL: Absent Without Leave

Basic Training: First 6 weeks of training for the military

Battalion: 3 to 5 companies of the infantry

Body Bag: A plastic bag to transport dead bodies

Body Count: Number of deaths of the enemy

Bush: The jungle or in the field

Cherry: A soldier yet to be in combat

Chinook: Transport helicopter

Chopper: Helicopter

CIB: Combat Infantry Badge

C.O.: Commanding Officer

Company: 3 to 4 platoons 100 to 150 soldiers

Dust-Off: Medical evacuation by helicopter

Flower Land: Vietnam

Gook/Dinks: Derogatory term for the Vietnamese

Grunt: An Infantryman

Hooch: Vietnamese hut or shelter

Hump: Waking with full load through the jungle

Incoming: Artillery/Mortars coming in your direction

In Country: Vietnam

Lifers: Career military personnel

L.P.: Listening Post soldiers outside perimeter for observation

L.Z.: Landing Zone for helicopters

MamaSan: Vietnamese woman

Medevac: Medical evacuation

Monsoon: Heavy rains, usually seasonal

M.P.: Military Police

Napalm: A jellied petroleum substance which burns fiercely

NVA: North Vietnamese Army

Piaster: South Vietnamese money

Platoon: 35 to 50 men

Pointman/Point: Forward man on a combat patrol

Rear Area: A secure base usually away from the jungle

Round: A bullet or ammunition

RPG: A Russian made Rocket Propelled Grenade

R & R: Rest and Recuperation away from the action

RTO: Radio Transmitter Operator

Short: A tour of duty close to completion

Stars and Stripes: The U.S. Military newspaper

The World: The United States

Tracer: A round of ammo chemically treated to glow when fired

VC: Viet Cong, South Vietnamese Communist, enemy soldiers

A Walk Through the Valley of Death

The End

Made in the USA
Columbia, SC
08 June 2023

077e6d97-5b65-4e7f-b6e7-a3e2472b14c3R01